OCT _ _ 2014

Social Exclusion, Power, and Video Game Play

D1293453

Social Exclusion, Power, and Video Game Play

*New Research in
Digital Media and Technology*

Edited by David G. Embrick,
J. Talmadge Wright, and Andras Lukacs

LEXINGTON BOOKS
Lanham • Boulder • New York • Toronto • Plymouth, UK

CHATHAM-KENT PUBLIC LIBRARY

Published by Lexington Books
A wholly owned subsidary of Rowman & Littlefield
4501 Forbes Boulevard, Suite 200, Lanham, Maryland 20706
www.rowman.com

10 Thornbury Road, Plymouth PL6 7PP, United Kingdom

Copyright © 2012 by Lexington Books
First paperback edition 2014

All rights reserved. No part of this book may be reproduced in any form or by any
electronic or mechanical means, including information storage and retrieval systems,
without written permission from the publisher, except by a reviewer who may quote
passages in a review.

British Library Cataloguing in Publication Information Available

Library of Congress Cataloging-in-Publication Data
The hardback edition of this book was previously cataloged by the Library of Congress as
follows:

Social exclusion, power and video game play : new research in digital media and
technology.
p.cm.
Includes bibliographical references and index.
1.Video games—Social aspects. 2. Fantasy games—Social aspects. 3. Role playing—
Social aspects. 4. Leisure—Social aspects. I. Embrick, David G. II. Wright, J. Talmadge.
III. Lukács, András.
GV1469.34.S63S64 2012
794.8—dc23
2011047269

ISBN: 978-0-7391-3860-1 (cloth : alk. paper)
ISBN: 978-0-7391-3861-8 (pbk. : alk. paper)
ISBN: 978-0-7391-3862-5 (electronic)

♾™ The paper used in this publication meets the minimum requirements of American
National Standard for Information Sciences—Permanence of Paper for Printed Library
Materials, ANSI/NISO Z39.48-1992.

Printed in the United States of America

To future critical sociologists and game researchers

Contents

Acknowledgments

This book represents the collaborative efforts of many scholars from multi-disciplinary backgrounds, perspectives, and interests. As we noted in our previous book, *Utopic Dreams and Apocalyptic Fantasies: Critical Approaches to Researching Video Game Play*, we feel that there are very few collections out there like ours and it is only because of these researchers that we are able to put together this book. It has been our pleasure reading these papers and working with our contributors who are deeply energized about this field and their cutting edge research. Therefore, with much appreciation and fanfare, we would like to sincerely thank our contributors for giving us this opportunity to highlight such great work. From the bottom of our hearts, thank you for giving us the privilege to learn about your work. We would also like to acknowledge our friends and colleagues in the Sociology Department at Loyola University, Chicago, for their unwavering support, feedback, and tolerance as we explored the sociology of virtual realms and occasionally shared our experiences with (sometimes too much) overt enthusiasm. Most notably, our department chair, Rhys Williams, deserves special recognition for believing in us and in the merits of this project. Your support means a great deal to us and you are well appreciated. Thanks again to Toby Dye and the staff at Loyola's Institutional Review Board for continuing to work with us closely on human-subject issues and for being open to developing new methods of researching virtual play. Thanks also to Gary Alan Fine from Northwestern University, Bonnie Nardi and Tom Boellstroff from the University of California, Irvine, and Thomas Malaby from the University of Wisconsin, Milwaukee, for all of your invaluable feedback on this project. We would like to give a special round of thanks to the great folks at Lexington Books and our editors, Jana Hodges-Kluck and Michael Sisskin. Jana, you truly are one of a kind. We could not have asked for a more patient person to work with as we struggled with the book chapters, consent forms, and our own internal time management struggles. Thank you for your vision, patience, and editorial assistance. As we struggle to legitimate social science in virtual realms, we owe a great deal of gratitude to scholars from various subfields with the intellectual curiosity to learn about our work and help us keep analytical distance from our object of study. Finally, we would like to thank our loving families for always being there, no matter how much we lost ourselves in the virtual world.

Introduction

David G. Embrick, J. Talmadge Wright, and Andras Lukacs

Multimedia entertainment and more specifically video game play, in its various forms, have increasingly become central components of peoples' lives. Although this is most notable in Western industrialized countries, more and more digital games are making significant waves on a global level. Indeed, in the U.S. alone, video games sales exceeded $21.3 billion dollars in 2008,[1] surpassing the total sales of the movie and music business and coming in second only after book sales.[2] According to MSN Money,[3] the predicted global worth of the gaming market in 2011 will be close to $49 billion dollars with the highest growth rate taking place in the Asia-Pacific region. Within the umbrella category of video games, massive multi-player on-line role playing games (henceforth MMORPGs) have enabled players of all calibers to create and recreate real-time social experiences with other players in a wide variety of virtual environments that range from the most fantastical to spaces that replicate the real life environments of the players themselves. Beginning with early games like *Ashron's Call, Ultima Online and Everquest*, more recent games like *World of Warcraft* currently command over 11 million subscribers to their virtual landscape. And this is only the tip of the iceberg; MMORPGs are growing at a phenomenal rate as players are able to access such games in a wide variety of interfaces that include not only the personal computer, but also through entertainment consoles such as the Playstation 3, X-Box 360, and Nintendo's Wii, and increasingly via smartphones (e.g., Motorola Android, Apple Iphone), smart tablets (e.g., Apple Ipad and Ipad2, Motorola Xoom, Samsung Galaxy), and cable television (e.g., Verizon's FIOS, AT&T's Uverse).

The expanding role of new forms of electronic entertainment such as digital games allow us to communicate, socialize, reflect, dissent, work, and play in new ways that were never dreamed of in the past. Indeed, the ability to traverse halfway across the world in mere seconds in order to conduct business or engage in idle conversation is a relatively new concept. Increasingly, video games and other forms of digital media take up more of our daily lives. For instance, according to market researcher NPD Group, the average number of hours spent playing games online has increased from 7.3 hours in 2009 to over 8 hours in 2010.[4] A study conducted by the Nielsen Company suggests that not only are people spending more time playing games, the number of subscribers to video game rentals services such as GameFly has dramatically increased.[5] Although such studies also conclude that part of the reason for increased sales is due to the economic depression, other research (see chapter 7 by Massanari) suggests, as least partly, that new innovations in game research and development (e.g., Nintendo Wii) have enticed more women and elderly to join the digital age revolution.

Additionally, the use of electronic entertainment in shaping relationships between people and altering the practices of everyday life has called into question traditional notions of work, play, race and ethnic relations, gender and power, for example. Further, it has also exposed generational fault lines and increased the number of venues for both youth and adults to engage in self-expression. As we have noted in a previous book, *Utopic Dreams and Apocalyptic Fantasies*, in the virtual realms of these new electronic environments, fears of violent representation mingle with myths of isolated game players. As an amplification of our contemporary fantasies, interactive media provides a virtual battleground for creating, reproducing, and sometimes even transgressing our everyday prejudices and assumptions.

SOCIAL EXCLUSION, POWER, AND LIBERATORY FANTASIES

Our contribution in this edited book is to address questions prompted by the expansion of new digital media and specifically, questions of how power is produced or reproduced by publishers, gamers, or even social media; how social exclusion (e.g., race, class, gender, etc.) in the virtual environments are reproduced from the real world; and how actors are able to use new media to transcend their fears, anxieties, prejudices, and assumptions. The articles presented by our contributors in this volume represent cutting-edge research in the area of critical game play, and it is our hope to draw attention to the need for more studies that are both sociological and critical.

We divide this book into three major sections that address major issues of social exclusion, power, and liberatory fantasies in virtual play. The first section, "Social-Psychological Implications of Virtual Play," highlights recent research that examines how the virtual realms of MMORPGs and other games shape emotion and influence social interactions between players within the game. Section two features studies that entertain questions on the marketing of race and gender stereotypes in video games and how (and if) traditional forms of social inequality are reproduced or contested in virtual realms. Finally, section three offers insight on game fans and virtual play. Specifically, the contributions in this section explore the relationship between software developers and game fans. We outline in detail each of these sections below.

SOCIAL-PSYCHOLOGICAL IMPLICATIONS OF VIRTUAL PLAY

Understanding the nature of modern digital play and the attraction of virtual realms has often been colored by a "media effects" literature. Research in this area is more often than not guilty of using overly simplistic models of human behavior which have contributed to mystification of the play process and stereotyping game players as "isolated," and/or "alienated." Much of the social science research, in fact, indicates quite the opposite with regard to game players and game consumers. This section highlights the institutional, political, and psychological sources that shape this distorted image of modern play. In particular, we are mindful of how our use of narrative and language shape our understanding of the pleasures (and pains) game players take away from their activity (e.g., addiction, social isolation, etc.). Further, how do virtual realms of MMORPGs shape emotion and influence social interactions between players within the game.

It is well known by game players that trusting others in online games is not a simple process, but, rather one which is continuously negotiated with others. How does such trust generate new forms of social solidarity and what are the activities which work to sustain such solidarity (e.g., guild or social network formation within games)? Also, how game players negotiate time between their off line families and the virtual realm is an indication of how the public comes to perceive the pleasure or threat of virtual realms. MMORPG guild structures also work to provide leadership skills to players, both young and old, providing new opportunities to break down age/competence distinctions which often trap adolescents in a prefigured social life. How do these new virtual realms break down traditional age barriers through play? And, finally, how exactly do game rules and the overall structure of virtual realms either help or hinder cooperation and teamwork.

We begin this section with Bertozzi's chapter, "Marking the Territory: *Grand Theft Auto IV* as a Playground for Masculinity." According to Bertozzi, games such as *Grand Theft Auto* allow males a virtual opportunity to explore traditional concepts of manliness and masculinity. Such games are constructed in a way that epitomizes what it means to be a man in today's society (e.g., bravery, courage, violence, aggressiveness, heterosexual dominance, etc.). And as a result of legal and cultural changes that discourage at least extreme forms of masculinity, games such as *Grand Theft Auto* not only allow males (and females) an opportunity to explore and overstep the boundaries, it rewards players for engaging in such extreme behaviors. Thus, according to Bertozzi, games such as *Grand Theft Auto* provide a virtual playground for players wishing to engage in a wide range of behaviors that are socially unacceptable.

The next chapter by Erkenbrack, "Discursive Engagements in *World of Warcraft*: A Semiotic Analysis of Player Relationships," addresses the common misconception of online games as immaterial. That is, ErkenBrack argues that virtual multi-player games such as *World of Warcraft* need to be closely examined in order to establish the "real" relationship between a player and the game—as a self-conscious reality for the players rather than a made up fantasy.

In chapter 3, "The Intermediate Ego: The Location of the Mind at Play," Vanessa Long addresses some of the structural issues related to virtual games, specifically those structures surrounding the user, the screen, and the avatar. According to Long, the idea of the disposable ego is important for understanding the point of interaction of the player with the video game space. She argues that players are able to navigate from one identity to another depending on their relation to the game and the computer screen. The intermediate ego serves to allow the player to engage in a process of becoming something which is not achievable, yet at the moment of play, players identify themselves outside of their true selves. Understanding how players operate freely between the game, the computer screen, and real life provides us with clues as to how we, as players, interact with games and what experiences we are able to take away from games given our interactions.

Finally, J. Talmadge Wright examines the diverse types of social relationships expressed in *World of Warcraft* in chapter 4 titled, "Producing Place and Play in Virtual Game Spaces." In this essay he examines a range of conversations, chat associations, and in-game behaviors which point to larger questions of social status, performance, gender, and age inequality, as well as demonstrating the fluid character of virtual performances with "real" life admissions.

SOCIAL INEQUALITIES IN VIDEO GAME SPACES: RACE, GENDER, AND VIRTUAL PLAY

There is a tendency to dismiss online virtual programs such as *World of Warcraft* and *Everquest* as "games" that people play either for fun, or to escape the realities of everyday life. What many of us fail to recognize is that these communities, although virtual, are extensions of the society in which we reside. That is, while these programs represent another social location from which people can communicate and interact with one another on a global scale, they are also virtual reproductions of racial and gender structures in our "real" society.

It is true that virtual realms and video games in general often make use of gender, race, class, and age stereotypes both in the marketing of the product as well as in game content. In the second section of our book, the authors examine how such stereotypes may be used or manipulated by both the game fans, as well as the developers. The question is asked: how does the playing of MMORPGs and other games in virtual realms either reproduce traditional forms of social inequality or contest those inequalities, or do they do both at the same time? Further, how do game chat and game behavior conform to or violate traditional notions of social stereotypes?

Since critical race studies, queer theory, and feminist studies all point to the significance of how race, class, sexuality, and gender are constructed on an everyday basis through everyday performances, we will examine how those constructions play out in fan behavior as well as game design. Finally, this section will also examine whose politics are represented in virtual realms, especially MMORPGs. Whose politics are promoted? Whose politics are repressed? Politics is often associated with a simple understanding of political parties. But, politics in the manner we are using here is about much broader issues of power and who gets to be seen and heard and who does not.

Pierre Bourdieu's (1977) concept of habitus provides insight into the nature of these practices and boundaries. For Bourdieu, habitus refers to socially acquired tendencies or predispositions which serve as a "matrix of perceptions, appreciations, and actions," causing individuals to view the world in a particular way (83). Habitus does not point to individual character or morality, but to the deep cultural conditioning that reproduces and legitimates social formations. Although individuals possess unique ideas and experiences, they tend to act predictably because they reside in the same social niche with others who are affected by similar rituals, belief systems, and interests. While the habitus does not determine action, it orients action. Thus, people observe and participate in social closure but tend not to see it as problematic as it resonates with their habitus. The habitus helps normalize and legitimate social closure.

Around this time, new media scholars suggested the relationship between the "real world" and the virtual was far more nuanced than these early utopian discourses suggest. Race, gender, class, sexuality, and (dis)ability did not simply disappear when interacting online; rather, these realities of our everyday world shaped (and were shaped by) our entry into cyberspace. Indeed, the very interface through which we engaged virtual/computerized environments often made us subjects to a complex network of power relations. In cyberspace, identities could be more fluid than in non-virtual spaces; however, the social-cultural structures that shaped our experiences in the everyday world did not cease to shape them online. Instead, our interactions were shaped by the choices designers and producers make when creating a technological artifact–and these choices were further influenced by professional design practices, organizational politics, and economics.

In chapter 5, Jessie Daniels and Nick LaLone examine the complex ways in which systemic racism is implicated in video gaming culture. According to the authors, neither research on video gaming nor research in the area of race has taken seriously the challenge of exploring the intersection of racism in gaming. They note that overt and covert forms of racism continue to exist in online games.

David Dietrich, in the next chapter, examines one way in which systemic racism is pervasive in the gaming world through avatar creation. One of the unique features central to electronic games and especially in MMORPGs is the direct interaction of real world players with other players as well as the computer generated non-role playing characters within any fictional gaming environment. The creation of an avatar allows real world players to create an identity from which to interact with other players or the environment. Dietrich argues that the process of avatar creation often reinforces ideas of whiteness being "normal." Further, the extreme limitations of creating avatars that are non-white forces players to accept the normativity of whiteness in these gaming environments.

Like race, gender stereotypes, both passive and overt, limit the ways in which women interact and are viewed in the gaming world. As Adrienne L. Massanari in chapter 7 argues, the majority of video games and video game consoles in the past forty years have catered to the young male crowd in both design and advertisement. In her chapter, she examines the ways in which the Wii challenges and reinforces the images that have become socially accepted in our society and which convey "appropriate" female game play.

In chapter 8, Joel Ritsema and Bhoomi K. Thakore specifically examine the white heterosexual masculinity framework that is privileged in MMORPGs. They argue that many virtual environment spaces are racialized and therefore offer players what Joe R. Feagin has labeled "sincere fictions of whiteness." These fictions help to keep whiteness normalized in a racialized social system. The authors call for more research that would contribute to the ongoing and relatively new dialogue on race and MMORPGs.

Last, Zek Cypress Valkyrie argues in chapter 9 that the current gameplay approach of many MMORPGs do not allow for true gender flexibility. Valkyrie notes that players who are able to "prove" that their real world genders are women have different gaming experiences in comparison to players whose real world genders are thought to be men. Women gamers were afforded nicer treatment by male players that included in-game assistance, money, and other game perks. However, such treatment was often accompanied by various levels of sexual harassment and gendered perceptions of gaming ability. According to Valkyrie, the enforcement of gender transparency in MMORPGs reproduces normative standards of femininity and masculinity found in the real world.

GAME FANS SPEAK OUT

While game fans often speak out in online blogs, Web sites, and feedback to industry developers through fan magazines, the authors in this section wish to move beyond the occasional story to a systematic understanding that game fans have of the pleasure they find in virtual realms. Hence, we finish this book by including chapters from authors who have worked with fans closely, employing social science methods to explore deeply fan pleasure and virtual realm play. Our next contributors explore the relationship between the software developers of MMORPGs and other online games and game fans. It is well known that game fans use game resources for producing more than simply another mod of a game. Such developments can range from political protests within virtual realms to art pieces and film (e.g. Machinima). This raises the broader question that the authors will take up, how should we understand modern fandom in the virtual world? Further, what is the assessment of the significance of play within virtual realms–the ability to alter real world economic and political power through the networks and forms of solidarity established in the virtual realm.

Sean C. Duncan starts out this section of our book by exploring the means by which "design thinking" is evinced by the interactions between *World of Warcraft* players and Blizzard game designers. Using discourse and content analyses, Duncan examines the discursive practices that participants employ in engaging with other participants around a particular game. In focusing on the design activities of participants within the online forums, Duncan questions the manner in which players argue for redesigns of the game's rules and mechanics and how participants discuss gaps in the conceptions of the spaces' purpose between game players and game designers. Duncan concludes by noting that scholars need to develop better ways of understanding the ways that participants' framing of their activities clash with or are shaped by game producers. There is value in understanding the interactions between game players and designers, especially in terms of today's fast changing technologies and burgeoning and specific consumer demands.

In chapter 11, Mia Consalvo focuses on one particular subset of videogame players which she labels "Western Otaku." These players particularly enjoy finding and playing Japanese games, often going to extremes to find earlier Japanese versions before Western or English versions are released. Using in-depth interviews with Western Otaku players, Consalvo draws a deeper understanding of who these players are and what, in particular, draws them to Japanese specific games. Further, Consalvo explores how transnational fandom operates in the realm of videogame culture and how globalization has changed or altered players' game play experience in the current climate.

Finally, R. M. Milner in chapter 12 argues for a new paradigm—which he calls the 'New Organizational' paradigm—to understand how productive consumers and the producers of media texts might interact with one another in our age of digital revolution. According to Milner, fans have long been thought of as 'textual poachers,' players who would traverse the digital landscape and try to adapt it to their various needs as best they could. Such mobility and ingenuity by fans are often in conflict with game producers whose ideal software products are those that they have full control over their productions. The result is confusion and miscommunication over ownership: specifically, how do fans perceive their level of ownership in the games they purchase and play? Further, how do fans and producers negotiate the issue of ownership in a global era laced with increased interaction between the two groups? Using discourse analyses on the official *Fallout 3* forum, Milner concludes that game fans are productive consumers who criticize and create and take part in a "gift economy" that operates interdependence with what game designers put out.

FROM THEORY AND METHODOLOGY TO ENGAGING IN CRITICAL RESEARCH ON VIDEO GAME PLAY

To conclude, we hope that the critical issues concerning modern play, video games, and game research presented here will prompt others to plunge further into this line of research. Given the importance of virtual worlds of play in modern society and the connection between those worlds of fantasy and consumer capitalism, it is important that we work to develop a critical perspective which keeps in focus the significance of social justice, domination, and liberation. Like radio, television, and movies, these newest technologies of representation are here to stay and rather than singing their praises or conversely dreading their corrupting influence, we would be best served by deploying the best of social science techniques and concepts and work to reveal how this new type of play can both free us in fantasy while also trapping us in conventional narratives.

NOTES

1. See Celebritynetworth.com for more information on total video game sales in the U.S. in 2008; accessed at http://www.celebritynetworth.com.

2. Total combined book sales in the U.S. for 2008 were roughly double that of video game sales ($40 billion dollars). Note that this figure does not account for the recent e-book explosion and sales of digital book interfaces such as Amazon's Kindle, Barnes & Noble's Nook and Apple's iPad.

3. Reuters, "Video-game Sales Overtaking Music" (MSN, 2007), accessed at http://articles.moneycentral.msn.com/Investing/Extra/VideoGameSalesOvertakingMusic.aspx.

4. Dean Takahashi, "Time Spent Playing Games Keeps Going Up" (GamesBeat: Interpreting Innovations, March 2, 2010), accessed at http://games.venturebeat.com/2010/03/02/time-spent-playing-video-games-keeps-going-up/.

5. 1Up.com, "Study Shows Time Spent Playing Games is Up During Recession," accessed at http://www.1up.com/do/newsStory?cId=3175102.

Part I

Social-Psychological Implications of Virtual Play

Chapter One

Marking the Territory

Grand Theft Auto IV *as a Playground for Masculinity*

Elena Bertozzi

In his book *A Primate's Memoir*, Robert Sapolsky examines the quality of life of males in a pack of baboons on the African savannah.[1] It is commonly perceived that when these packs are dominated by an alpha male, he has the luxury of sexing any female of the group he wants, whenever he wants. Such an interpretation has often been applied to human behavior, in that it explains male desires for multiple partners and justifies ideas of male superiority.[2]

Sapolsky found that the reality is quite different. The lives of males in such societal structures, even those of alpha males, are often much more complicated than thought. The structure of male hierarchy is fluid. There are always males looking to replace the alpha male. Thus, males have to spend much of their lives fighting to defend whatever position they have achieved. Saplolsky's research specifically measured stress levels in the males. He found that male baboons suffer from a great deal of stress, spend a lot of time fighting, and often die from stress and violence-related injuries. Welcome to the world of *Grand Theft Auto IV* (henceforth referred to as *GTA*).[3]

The goal of this chapter is to consider the specific kinds of pleasures afforded by *GTA* and specifically the latest game. The *GTA* games are notorious for many reasons. They are considered so violent and depraved that some countries have gone so far as to ban their sale.[4] It remains, however, one of the most successful game franchises of all time, having sold more than 70 million copies.[5] If the game were just violent, it would simply be one of a large number of other similar games.

GTA is special. It provides male players with specific and finely tuned opportunities to perform idealized masculinities. Players have opportunities to be the kind of male that is presented as the hero in myriad films, television shows, and advertisements; a male they are highly unlikely to actually be outside of the game. It affords a chance to achieve idealized masculinities and make them real, if only in a virtual setting.

Manhood, the nature of masculinity, and social constructs of gender vs. biological determinism are topics of much debate in the 21st century. The discussion's heated nature is reflected in the changing structure of academia itself, as more Women's Studies departments change their names to Gender Studies or are inclusive of Men's Studies. Modern men are encouraged to be involved in parenting, show their emotions, have caring and nurturing relationships with others, and live lives different from those available to males a generation ago. These kinds of men are the heroes of what are called "chick flicks"—or films whose primary audience is female. Stars like Hugh Grant specialize in playing soft, emotionally conflicted, and often vulnerable males. These men often seem to be trapped in a kind of Peter Pan universe where it is difficult for them to behave like adults. Seth Rogen's films such as the *40 Year-Old Virgin*, *Knocked Up*, and *Pineapple Express* also portray males in non-traditional ways. Rogen's heroes are often not especially attractive or physically strong. They also engage in many adolescent behaviors like dressing up, playing practical jokes, and obsessing about the sex that they are not having.

These two genres of films both reflect and define new ways of being male in the 21st century. The chick flick genre defines what women supposedly would like men to be, while Rogen-type films purportedly show men as they really are. Researchers in Men's Studies have documented the fact that the conceptions of what it means to be male in modern Western society are increasingly flexible.[6] Culturally acceptable ways of being male are much broader and more accepting of ambiguity than they have been in the past.[7] This new freedom does not come without cost. Some have argued that the shifts in gender identities make it difficult for men to understand how they should be male and what kind of a life will lead to feelings of fulfillment and satisfaction.[8]

Although modern life provides men with many more flexible interpretations of masculinity than were available in the past, hegemonic models of the ideal male remain strong. Popular culture continues to propagate images and stories of idealized males who embody traditional stereotypes of masculinity. Films, television shows, and advertisements portray men who are: hard, aggressive, competitive, violent, willing to sacrifice themselves for honor, desirous of rescuing females, unemotional, detached, etc. *GTA* allows players to be a kind of idealized male for a period of time within a world with few ambiguities. The player plays as a man who has to earn respect from other men through a series of violent acts. He has to acquire weapons and use them without compunction. He is loyal to his allies and rescues them when needed.

He establishes relationships with women based on their need for his strength and determination–one betrays him, the other is killed and her death avenged.

In this chapter, I detail the pleasure to be found in enacting various tropes of hegemonic masculinity without actually having to suffer consequences in the real world. *GTA* has achieved its popular status, in part, because of the vast disconnect between the way many say they want their men to be and the way idealized masculinity is portrayed in our culture. *GTA* is a way for male players to bridge that gap.

As Connell points out, what it means to be male is constantly changing both in terms of what is appropriate and how it relates to economic class. Male roles shift in response to changes in the world around them:

> The new information technology requires much sedentary keyboard work, which was initially classified as women's work (key-punch operators). The marketing of personal computers, however, has redefined some of this work as an arena of competition and power–masculine, technical, but not working-class. These revised meanings are promoted in the text and graphics of computer magazines, in manufacturers' advertising that emphasizes "power" (Apple Computer named its laptop the "PowerBook") and in the booming industry of violent computer video games. Middle class male bodies, separated by an old class division from physical force, now find their powers spectacularly amplified in the man/machine systems (the gendered language is entirely appropriate) of modern cybernetics. [9]

This suggests that not only are computer-mediated worlds like *GTA* enjoyable playgrounds for modern males to explore, but the act of playing "hard" masculinities on a computer concomitantly reinforces the link between masculinity and technological mastery. The ability to show competence in such worlds is important to establishing and enforcing masculinity outside the game as well.

2008 MOVIE HEROES

Men's roles in the Western world have changed dramatically over the past three decades. The United States and Europe have moved from cultures in which a large percentage of men (especially working class men) had jobs that involved hard physical labor to jobs consisting of activities such as: working in offices, driving vehicles, operating computer technology, managing people and things, and working in the service industry. [10] There is little need for brute physical strength in such jobs and many of them were considered jobs for females before the shift from agriculture and manufacturing in Western economies.

Despite the fact that the lives and roles of men have changed dramatically, cultural myths about the ideal heroic male remain largely unchanged. The ten top grossing English-language films of 2008 were: *The Dark Knight, Indiana Jones and the Kingdom of the Crystal Skull, Kung Fu Panda, Hancock, Mamma Mia!, Madagascar: Escape 2 Africa, Iron Man, Quantum of Solace, WALL-E,* and *The Chronicles of Narnia: Prince Caspian.*[11] If animated and children's films and the sole "chick flick" were eliminated, the only movies that remain are ones that include representations of idealized, heroic masculinity as dominant themes.

A man alone—In each story, there is a man who operates largely by himself. He has friends and associates who help and support him, but he is generally shown as isolated from others for some reason. He is often a rebel, and he actively resists traditional authority structures like the police or the military.

Evil threatens—There are dark forces working against the man. He is aware of the dangers they entail, but he is intelligent and sensitive enough to appreciate the danger. This man is also brave and of good character. He is somehow selected to be the person who must combat forces of evil.

Strength and intelligence—The man is clearly fit and strong—though not overwhelmingly so. He is able to combine his physical strength with intelligence and mastery of the technology available to him (often actual superpowers) to fight against apparently impossible odds.

Women need him—In the story, there is a woman who is a witness to his acts of heroism and often needs to be rescued. It is often love for the woman or desire for the adoration or reciprocated love that motivates the man on a mission.

Technological mastery—All of the 007 movies involve the hero manipulating a range of technological gadgets with ease and dexterity. Technology is literally bent to 007's will, as anything he needs is customized for him. *Quantum of Solace* is no exception. The film *Iron Man* takes this conceit even further.[12] Not only is the hero a CEO of his company, thus demonstrating that he can manage money too, but he is a computer programmer and able to create top-level technology from shards of metal he finds lying around in an Afghan cave.

Hard work—His is not a soft desk job (though in several of the super hero movies, he actually does a desk job as a cover). He embodies the attributes of passion, action, competitiveness, justice, and especially the willingness to use violence to accomplish his goals. He is also required to physically suffer. He will be injured or terribly hurt, but perseveres regardless.

The storyline of *GTA* has many parallels to the story arc of these films.[13]

A man alone—Niko Bellic, the protagonist of *GTA*, arrives as a poor immigrant to Liberty City in the United States. His only contact is his brother Roman Bellic, who had told him lies about success in the "promised land."

Niko discovers that his brother is deeply in debt to men who are willing to kill to collect. Niko has no friends. He has gambled his life on his ability to make things work out in his new home. Embarking on the game is a way of forming an alliance; the player takes on Niko's mission as his own.

Evil threatens—There are some rotten characters in Liberty City. Rival gangs control different parts of the city, and newcomers are not welcome. The "bad guys" have lots of weapons, and Niko has to be clever and resourceful to combat them. In fact, Niko is working for a group of the bad guys. However, Niko is not like the others. He shows himself to be separate and unique due to his motivations and overarching desire for a better life.

Strength and intelligence—Niko moves through the world like a man who matters. He is not overwhelmingly muscular or hyper-masculine in appearance, but he moves with authority and weight. When next to other characters in the game, it is obvious that he is tall and strong. His perpetual 5 o'clock shadow further accentuates his toughness. Although his demonstrated intelligence is limited by the player's ability, his intelligence is amplified when compared to the stupidity of his brother Roman. Niko's dialogue is articulate, ironic, and sometimes funny. He dryly assesses situations and remains calm even when others around him become agitated.

Women need him—GTA is different from earlier versions in that there are many fewer prostitutes on the streets, particularly during the day.[14] This may be a response to external political pressure: Numerous critics have pointed out how players can have sex with prostitutes, kill them, and take back the money they had paid. This point was endlessly cited as proof of turpitude in earlier games of the *GTA* series,[15] but the change gives the newer version a different feeling.

Although few significant female characters were included in storylines of previous games, a large number of the women who appeared in the game wore skimpy clothes and looked to be sex workers. *GTA* is quite different in this respect. Players meet Roman's assistant, Mallorie, early in the story. She works in the office and wears ordinary clothing. People flirt with her, but they do not treat her like a prostitute. Niko is also introduced to her friend Michelle, and they actually date each other. Sadly, players later learn that Michelle is an undercover cop and her interest in Niko is not because she likes him. Later though, Niko meets and comes to care for Kate. The need to avenge her murder is one of the motivations that moves the storyline through to the ending.

Hard work—Niko's job is difficult. He routinely has to beat people to death with his bare hands or with an array of weapons. He must drive fast and ably, run up and down stairs, jump across rooftops, and do a variety of other physically exhausting tasks. The level of his physical strength is amplified by the way he moves and the way he is shown in contrast to his pudgy, ineffectual brother Roman.

Technological mastery—Niko must demonstrate mastery over a range of technologies to win the game. He must be able to drive any of the vehicles necessary for missions: cars, motorcycles, helicopters, boats, etc. At various points in *GTA*, he is also required to use and understand: cellphones, computers, email and chat softwares, and a vast array of weapons. Similar levels of technical competence are also required of players.

In order to be able to play the game at all, players have to be able to understand and use the interfaces of the PlayStation 3 and Microsoft Xbox 360. The controls for these consoles have multiple buttons, sticks, and bumpers that must be pressed, tilted, and moved in a very complex series of movements in order to accomplish in-game tasks. Without mastery of these controls, it becomes difficult for a non-expert player to even approach the *GTA* games.

In short, much of the *GTA* series' consumer popularity is due to the creators' understanding of the power and appeal to stereotypes and mythologies of heroic masculinity, combined with beautifully technically-executed games. Players are encouraged to break taboos,[16] but they are also enabled to go beyond the boundaries of heroic masculinity as portrayed in more traditional media. "Niko goes so far over the edge sometimes that he is not a charismatic anti-hero, he is a homicidal terrorist."[17] This too though, allows players to be "real men" in that he (or she) can follow the urge to be overcome by blood lust and simply kill anything in sight. Such features of the environment, avatars, and gameplay combine to create an extremely engaging and immersive environment. The following sections outline specifically the ways in which players can act out particular types of manhood within the confines of Liberty City—the freedom . . . or the constraints of battling to the death in the human equivalent of the baboons' savannah.

SPECIFIC MEANS BY WHICH MASCULINITY IS DEFINED AND EXPRESSED

Cars and Driving

Driving, particularly in an aggressive way, is closely associated with traditional views of masculinity.[18] Young males have long proved their mettle to themselves and others by driving "fast and furiously." A teen ritual documented in many popular movies is that of young males challenging their peers by revving engines and seeing who can drag race down some straightaway the fastest. The practice is deadly and thus the topic of many driver safety courses. Driving accidents are the single most likely way for a young male to die in the United States.[19]

As the title of the series clearly states, *GTA* is based on the player's ability to interact with cars. A fundamental premise of the series is that players can engage in a variety of taboo, illegal, and fun activities with cars. An enormous part of the pleasure of the game is the fact that it gives players opportunities to be bad[20] and to perform specific dangerous behaviors associated with testosterone-addled young males. In all the *GTA* games, players can approach any car, wrench open the door, throw the driver to the ground, and drive away. Rarely will the driver fight back, in which case it is necessary to use violence to get away.

GTA players can ignore the storyline and missions and simply spend hours exploring the city in different automobiles. Cars are scattered throughout the city, and many of them explicitly reflect existing real-life models.[21] Each has a unique driving style, with varying ability to handle speed and accidents. Even more, each car features different music with stereotypical radio stations that seem to match the original driver. They can drive over curbs, smash through light poles, run over pedestrians, emerge victorious from police chases, and provide a perfect staging point for several different kinds of weapons. *GTA* provides a environment in which players can enact almost any of the aggressively masculine car-related behaviors that are prohibited either by law[22] or public opinion in modern society. Nonetheless, such masculine behavior remains tantalizingly appealing due to continued glorification in popular films and television shows.

Prey, Predation, and Intelligence

One of the binary divisions in traditional definitions of masculinity and femininity is activity/passivity. Males are to act forcefully to seek out females and accomplish external goals, while females passively await male attention and focus most of their activity on improving their status as sex objects.[23] Games such as *GTA* provide many opportunities for males to be forceful and active, though very few of the activities enabled by the game are in any way constructive. One of the most interesting activities supplied by it and many other male-oriented games is the notion of predation. The thrill of the hunt and exquisite anxiety provoked by lingering threats of death maintain player interest and emotional investment throughout the game.

In his fascinating book, *The Better to Eat You With*, Berger[24] discusses the role of predation in evolutionary intelligence. He compares behaviors in populations of prey animals in situations where predators (wolves) were eliminated by humans, and then what occurred subsequently when predators were reintroduced. He found that moose that grew up without predation did not know that they ought to be afraid of wolves. They were easily and quickly killed by the reintroduced predators, suggesting that fear of wolves is a learned behavior, not one that is naturally encoded.

Berger notes that the behavior of prey populations was significantly different if they experienced predation. Populations that were preyed upon had to become smarter. Animals that failed to learn about the dangers of predators were quickly killed. The surviving animals were those who were aware of the threats and developed strategies to detect and counter such threats. They were the animals who managed to reproduce and subsequently teach their young to fear predators. Thus, the presence of predators acts as a selective force on prey populations, ensuring that the ones who survive are the ones who can learn and pass that knowledge on.

Wolves, predation, and the relationship between wolves and humans are topics considered in Jiang's fictionalized account of his time spent on the plains of Mongolia.[25] The nomadic tribes that live and raise sheep on these plains have a very complicated relationship with wolves. Sheep are preyed upon by wolves, but nomads hunt the wolves under certain circumstances. Jiang describes cultural effects of this relationship. The predator/prey relationships create ongoing competition that results in more intelligence among both groups. Each side is constantly seeking better, more strategic ways of beating the other, so they are constantly learning from one another. Jiang argues that this relationship is so important to the nomads that the wolves are seen as a kind of deity. They are seen as having mythical powers, which are both dangerous and beneficial to the nomads.

Modern society provides few opportunities of such predator/prey experiences for most people. In many parts of the United States, hunting remains a popular sport and one closely associated with hegemonic masculinity.[26] However entertaining it may be to hunt as predators, few modern males routinely experience being prey. One opportunity is presented by the increasingly popular sport of paintball which was played by over 5 million Americans in 2008.[27] Paintball is a simulation of a warzone which means that all players are simultaneously predator and prey. It is an emotionally intense, engaging game. Paintball is extremely physically demanding and requires specialized equipment and environments. *GTA* is much more accessible.

Like many other shooting games, the bulk of the missions in *GTA* involve hunting people down and killing them. In the process of doing so, Niko angers other actors in the game who are focused on killing him as well. The experience of hunting and being hunted is what raises the level of tension in the game and what maintains it over the long periods of time required to complete the game. The player experiences the feeling of becoming smarter as he survives predation and as he preys upon others.

GTA heightens the experience of being hunted. When Niko is wanted by the police, for example, a red circle surrounds him. The player has to figure out how to get Niko away from oncoming predators within a given period of time. The tension is augmented by the red color of the screen and the wailing of approaching sirens. Niko often has to race away through twisty streets making lots of fast turns to throw cops off his tail. If he fails, he can be shot while resisting arrest or he will wind up either in the hospital or back in front of the local precinct minus money and health. At this point, he must restart the mission from scratch.

The game utilizes the predator/prey environment as a way of making players feel intelligent. Players must outsmart aggressors to complete missions and survive, but are punished for wrong moves by succumbing to abler predators. Unlike succumbing to wolves on the Mongolian plains, when Niko is killed by a predator he "respawns." The game provides the pleasure of positive experiences without pain of the negatives. Successfully beating the game involves determining how the police and other predatory males will act and figuring out how to accomplish the mission by either avoiding or killing them. Surviving predation gives players feelings of accomplishment through incentives (e.g., increased health, money, and other in-game rewards).

Pride

GTA IV specifically reinforces traditional concepts of patriarchal masculinity. Pride is an essential component of this, and it is important for establishing the ability to lead and generating social capital. [28] Niko does not have many material assets beginning the game, though he accumulates them over time. What he does have is pride. This is expressed through gameplay as a motivational mechanism to encourage Niko to do what he is supposed to do, [29] and as a way of justifying the violence he carries out. It is a delicate balance because in order for the storyline to make sense, Niko has to obey the orders that come from above. Thus, he is technically a servant of more powerful men. However, the game utilizes the pride structure familiar to Mafia-type organizations and cultures in which blood feuds are the norm to justify Niko's behavior.

Any slight or expression of disrespect from one man to another in the game must be paid for with blood. Niko is routinely charged with killing someone because that person has offended someone above him. Traditional patriarchal values regarding women are used to motivate Niko on several occasions. At one point, Vlad takes an interest in Roman's assistant Mallorie (*Uncle Vlad* mission), and Niko is sent to talk to Vlad about leaving Mallorie alone. Of course this does not work and Niko must follow Vlad to the docks and kill him. Mallorie is not consulted.

In another mission, Mikhail decides that he does not like his daughter's biker boyfriend. Niko gets the contract to eliminate this problem (*No Love Lost* mission). He follows the boyfriend to a park where he is attacked by a group of bikers. Niko must kill them all to complete the mission. Not only is Niko enforcing paternal control over the daughter's love interest, but he is doing so while defeating traditionally hyper-masculine males. When Dwayne's girlfriend leaves him for another man, Dwayne hires Niko to kill the guy and Cherise. Later it is determined that Mikhail should not have ordered the killing of the biker boyfriend because he had high-level connections. Consequently, Niko is required to kill Mikhail too. This is the code of the street and how hierarchy is established and maintained.

MARKING THE TERRITORY

GTA, and many other games such as *Halo* and *Counterstrike,* require expert knowledge of the territory in order to succeed. A player's knowledge of the map(s) and understanding of the people and objects located within it is essential to winning the game. Such understanding allows game-play to be quicker and easier. For example, a player can go into the map and set waypoints for locations he (or she) would like to get to quickly, then steal a cab and get there within seconds using the in-cab navigation system.[30]

In *GTA*, the player must understand how to use the in-game map. Symbols on the radar indicate where to go and the locations of important people. Failure to understand this means that the player will simply wander aimlessly. *GTA* uses the cellphone as the interface to many of the important people and activities in the story. Over the phone, the player has to be able to make connections between what Niko is told to do and the colors and symbols that appear on the map icon in the corner of the screen.

Interestingly, several studies[31] posit that there are significant differences between male and female brains and how members of each sex orient themselves in environments.[32] Many couples have experienced this while driving in cars, having arguments about the best way to get to some desired destination. When the author of this study started playing *GTA*, she found it impossible to orient herself in the game without having the paper map that ships with the game. She spread it out beside her, and this helped her figure out where Niko was going in the game. Such a practice, however, was a source of hilarity to the young males she enrolled as expert assistants. Two of them said that they had never even seen the map because they always buy the games "used," while the third just shook his head in disbelief. They all disdained using the paper map and relied exclusively on in-game technology and personal experience for navigation within the game.[33]

When reflecting on this experience, it conjured an analogous parallel between the differences of taking male and female dogs for a walk. A male dog, particularly one that is un-neutered, cannot walk far without urinating, or marking the territory. He is also intensely interested in smelling the markings of other dogs. He frequently must be dragged away from this task to continue the walk. Female dogs appear to pay little attention to such signals. Failure to pay attention to territorial markings in *GTA* translates into certain death.

The game codifies marked territories through the use of "Safe Houses." Niko must have a place to rest and/or hide in between missions. Given that Liberty City is vast, he has to acquire Safe Houses in various locations across the map so that he can reach them during critical moments. He acquires Safe Houses in the process of completing missions. Niko is concretely rewarded for having made a hit or completing missions by acquiring new real estate that becomes marked as his.

Relationships

One feature of new forms of masculinity is an emphasis on the importance of relationships—meaningful human connections between males and females and male peers.[34] This can be seen in several of Apatow's films.[35] *The 40-Year-Old Virgin*, for example, features a man who has failed to have sex with a female. His friends get together to help him resolve this difficulty. The film demonstrates many different forms of relationships between males—generally in a positive and humorous light—while at the same time chronicling the difficulties of relationships with women. The film ends happily with the virgin finding, bedding, and marrying his true love.

Needless to say, *GTA* ensures that no such relationship business hinders Niko's progress through Liberty City. Niko solves the problem of relationships by lacking emotional sensitivities. If he wants sex, he can pick up a prostitute or go to a strip club and buy it. He does develop a kind of relationship with Michelle in that he drives her places and they go on a couple "dates," but little emotionally viable communication is exchanged between the two. If Niko neglects to contact her, he receives angry texts demanding attention. The only thing Niko has to do to prepare for these dates is to go out and purchase new clothing before picking her up. Interestingly, Niko also needs to be wearing a suit for the final missions in the game. This suggests that being well dressed is only partially about appealing to women, but also has a role in establishing status and appearing powerful to other men.

On one level though, the game appears to take relationships seriously. It codifies them and uses them to control some in-game events. Many relationships with other significant characters in the game go up and down based on Niko's choices. If Niko keeps his "Like" and "Respect" scores high enough in his relationship with Roman, for example, he can call him at any time and he will send a cab to take him wherever free of charge.[36] Thus, relationships are reduced to a kind of currency that can be spent as needed in the game.

Relationships with women lack depth. Niko has little reservation about using force, if he deems it appropriate. When Niko's job is to kidnap Gracie and she starts grabbing at the steering wheel, he slaps her (*I'll Take Her* mission). As in the earlier games, the avatar's ability to torture, humiliate, and kill prostitutes is extremely disturbing.[37] This, too, is directly related to traditional patriarchal views of women's role in society. Many cultures denigrate women who are viewed as promiscuous or who are sex workers; violence against them is often permitted or encouraged.[38] Late in the game, Niko meets and comes to care about Kate. Once again, little occurs to establish a relationship with her. She does serve as a motivator in the plotline, however. In one of the two endings, when she is shot to death at Roman's wedding, avenging her death increases the satisfaction of killing Pegorino. In the other ending, she leaves Niko—another betrayal, damn women! However, none of these relationships significantly affect gameplay or the player's experience of emotion in the game.

Relationships with men are even simpler. Niko has to help Roman because they are linked by blood, and that is all that matters in patriarchy. When it comes time to execute Vlad, Niko says, "You were the stupid one, Vladdy boy. Nobody fucks with my family."[39] Niko is constantly having to save Roman from dire situations in which he is being beaten or about to be killed by rivals. Roman's passive (i.e., feminine) dependence on Niko is emphasized by the way he is repeatedly referred to as "your fat cousin."

Niko accepts and executes the orders that come down from the bosses because in a pack of males, those below the top dogs obey them until it is time to kill them and occupy their position. When Niko is sent to "persuade" other men to do as they are told, this persuasion occurs in the form of beatings and physical threats. After the male has been subjugated to an adequate degree, he agrees to do Niko's bidding or is killed. Sometimes (as in the case of the *Weekend at Florian's* job) it is necessary to do both. If Niko does not kill him after he gets the information, the victim calls for help and makes life more difficult later.

Niko is always rewarded for completing his missions with cash and sometimes other benefits. The results of his success always benefit him alone. Sometimes men in the game will complain that Niko is working for someone else—usually an enemy of that particular mission contractor, but it is understood that this is business and not personal. Niko does what he has to do for money. In this regard, *GTA* is unlike other popular video games such as *Counterstrike* and *Halo*, which emphasize cooperation among groups of males. Cooperation is very often a smarter strategy than selfishness: "if people can arrive at a cooperative solution, any nonconstant sum game can, in principle, be converted to a win-win game."[40] *GTA* is not about optimizing group experiences or building relationships. It specializes in delivering the experience of the selfish, violent, aggressive, testosterone-driven male in unadulterated form.

Violence and Self-Sacrifice

As Messner et al. explored in their work on the "Televised Sports Manhood Formula," there are specific means through which manhood is defined in relationship to violence.[41] Public manhood can be described as a kind of performance:

> What is a Real Man? A Real Man is strong, tough, aggressive, and above all, a winner in what is still a Man's World. To be a winner he has to do what needs to be done. He must be willing to compromise his own long-term health by showing guts in the face of danger, by fighting other men when necessary, and by "playing hurt" when he is injured.[42]

Many other researchers have looked at this topic as well.[43] All the *GTA* games focus on permitting players to be extremely violent. Niko can roam the city and start fights with any passerby. He can drive recklessly and crash cars with abandon just because he wants to. As weapons and money are acquired through gameplay, the amount of mayhem players can achieve increases. The ability to deploy grenades, rocket launchers, and helicopters allows players to make some very big bangs.

Traditional masculinity often requires tests during which males must endure physical pain and inflict it on others. The ability to be a male is related specifically to the willing submission to violence and a corollary willingness to be violent toward others. This concept is delineated in the film *Fight Club*. In this film, modern males seek out other men with whom they can be "real men," and they do this in a world that they perceive as overwhelmingly feminized.[44] The depiction of violent manhood in this film specifies that:

1. it is a place separate from women
2. a man is required to be violent immediately upon joining the group
3. a man is required to submit to violence immediately to establish worth
4. participants experience the pleasure of watching violence occur around them.[45]

From the beginning, *GTA* establishes that it creates a virtual world for men, one where brothers take care of brothers. Women are largely irrelevant except as motivation for a job (protecting family honor, kidnapping), revenge (avenging deaths), or for apparently meaningless social encounters (bowling with Michelle). Earlier games in the series were populated by lots of prostitutes which further marginalized females as objects to be used for a specific purpose.

The game does not technically require violence if the player's only desire is to be an observer. A player can enter the game and just drive around and see the sights of Liberty City. Arguably, however, such a player is not playing the game because 90% of the game content is not available to a person who plays like this. In order to meet the cast of characters, experience the story, and have access to many parts of the map, the player must play and playing is being violent. Niko's job is to go out, kill people, and destroy things.

In the process of doing so, Niko is harmed and killed repeatedly. One of the advantages of playing video games though (unlike physical sports such as football and hockey where the pain is real), is that the avatar immediately reappears in the game, none the worse for wear. Niko has a health meter which must be watched to ensure that he has enough health to complete his missions. If he dies, either from violence or accident, he merely reappears at the hospital where he can set off to attempt the missions again. The avatar body can be repeatedly sacrificed—as the myth of manhood requires—without cost to the actual body of the player.

Interracial Conflict and Masculinity

The *GTA* series are notorious for racial stereotyping. A group representing Haitians sought to ban *GTA: Vice City* for its representation of Haitians.[46] Many of the storylines in earlier games in the series involved rival ethnic groups competing against one another for drugs, prostitution, gambling, and other sorts of crimes. The newer version is no exception. Roman and Niko, immigrants themselves, have to deal with American black gang members, Italian Mafiosi, Russian Mafia, and assorted others. (The latest game, released in spring 2009, is based in Chinatown.) From the beginning of the series, part of the pleasure of *GTA* has been the player's ability to feel powerful and behave violently towards ethnic groups that are considered dangerous.

Nakamura coined the term "identity tourism" to define the practice of players adopting another racial identity within a game and then enacting the stereotypes of that ethnicity.[47] Few players of *GTA* would feel comfortable actually walking the streets of dangerous urban ghettos controlled by any of the ethnic groups routinely represented in the *GTA* games, but inside the game the player is empowered in these environments. If Niko has the right weapons and enough health, he can blow away anyone. One example of this is the voice of a black drug dealer alternately pleading and taunting as Niko hunts him down and kills both him and his friends:

"Don't get so close to a brother . . . yo! Get off a brother's back! . . . I'm just trying to put food in my baby's mouth. . . . Motherfuck. . . . You're going to get the hell beat outa you boy. . . . You're going to get punished son. . . . The pain train's gonna get you bitch.You best not be coming in here. . . . Stay away from here . . . he's comin in. . . . You don't got us . . . ahhhhgg."

The caption describing this mission in the YouTube-published clip reads: "This one shows a mission where you follow a drug dealer to his crib, and then pop a cap in him and his homies."[48]

The ability to kill enemies seen as very dangerous goes a long way towards establishing rank and status in male power hierarchies.[49] This game allows players to perform this activity over and over again under different circumstances and varying degrees of difficulty. The hit often has to be outsmarted or outmaneuvered, not simply overpowered with force. Thus, this allows players to feel smarter and more powerful.

Cops

The role of the police in civil, democratic societies is to enforce the law and capture those who do not follow it. *GTA* provides players with the opportunity to defy authority and sometimes humiliate the police. A "real man" does not take orders from cops. This concept of masculinity that rebels against societal rules and strictures is codified in many popular films and repeated in the films cited above. In *Iron Man*, the hero has links to the military and other formal societal structures, but he operates outside of them and frequently disobeys orders from official sources. The *Indiana Jones* and *007* films are similar. Bond is working for a government protection agency, but he often breaks the rules and defies authority. Indiana Jones frequently scoffs at authority and finds his own way around hurdles.

In *GTA*, the act of specifically defying police authority is a central construct of the game. After committing a crime, the red ring covers the icon in the lower left hand corner of the screen and the voice of an officer can be heard ordering Niko to stop. To complete the mission, Niko needs to evade the police. This is accomplished by driving away as quickly as possible in explicit defiance of the officer. Escape often involves crashing cars, running over innocent bystanders, and shooting or killing as many policemen as possible.

Some missions specifically require the theft of a police car. Niko occupies the technology of authority and uses it against itself. In the *Portrait of a Killer* mission, Niko must access the computer in a police car to get information about some man he is supposed to kill. This occurs with even more sophistication in the *Smackdown* mission. Here, Niko uses the police car computer to find and track the victim, and positions the police car to help him. Then as he runs away from the crime scene, he can call 911 so that the police are keeping his pursuers busy while he flees.[50]

The player can also get a sniper rifle and locate Niko on top of a skyscraper with a high wanted rating. Then, he can attract police helicopters to the skyscraper by shooting at them. Using the rifle, Niko can pick off the police officers one at a time or shoot the pilot. If he shoots the pilot, he can watch the helicopter crash into the street while officers yell "Where do you think you are going to go? Put your hands behind your head!"[51]

Young males are often the targets of police attention. Rowdy males can get in trouble for underage drinking, driving in an unsafe manner, behaving badly at sporting events, college pranks, and a range of other activities. *GTA* provides players with a sort of payback experience. Everything that the player does in the game provokes the wrath of the police, but the player is able to thwart them at every turn. Successfully playing the game is an act of defiance against established authority.

Keeping Score

Traditional masculinity is often codified and expressed through ritualized and publicly displayed hierarchies. This is evidenced by the use of badges, patches, types of uniforms, and other ranking markers in the military, police, fraternities, and some organized sports. A participant in these organizations knows specifically what his rank is compared to others. This creates stability within the organization, but it also creates an orderly system through which those lower in the hierarchy can aspire to move up by achieving whatever is necessary to gain a higher position.

Another function of this display is bragging rights. A person who has obtained a particular status is able to display it and accrue the pleasure of implicitly and explicitly gloating about his achievements. The philosopher Johann Huizinga who wrote the seminal work on human play, *Homo Ludens*, explains the importance of boasting. "This boasting of one's own virtue as a form of contest slips over quite naturally into contumely of one's adversary, and this in its turn becomes a contest in its own right. It is remarkable how large a place these bragging and scoffing matches occupy in the most diverse civilizations."[52] He also posits that "the real motives [for war] are to be found less in the 'necessities' of economic expansion, etc., than in the pride and vainglory, the desire for prestige and all the pomps of superiority."[53]

Games that are produced for a primarily male audience exploit this desire for public vainglory in several ways. Within the game, *GTA* creates a clear and visible tracking system by which achievements are recorded and recognized. The main game has a series of missions that must be carried out, but there are side missions with separate lists of achievements and rewards.

Online game play and the ability to record and display individual game scores online have given players a much broader public. The X-Box and PlayStation 3 have made marketing choices that ably exploit players' desires to both show the scores they have obtained to the widest possible audience and to scoff at the score of others in online forums:

> On foot, by plane, sea, or car *GTA IV* is a blast to simple [sic] engage in an all out war for bragging rights and online cash that can be spent on your customizable character. Online isn't as smooth as it could be as it has a lot of glitches, however its way more than I expected and its 100% pure fun. Hardcore online gamers might not be satisfied, but most will be more than overjoyed to finally be able to lock and load with some friends online. [54]

Irony and Social Commentary

One of the most important aspects of *GTA*, and possibly what distinguishes it most from other video games is the irony that permeates the game. While the game allows players to indulge in the most extreme and implausibly testosterone-fueled activities, the game itself is mocking such behaviors. In-game television programs and advertisements, radio stations, and billboards provide a running satirical commentary on the state of civilization in general, and on the roles of males in particular. A veteran player of the game observes that:

> If you generally listen to the radio stations: *We Know the Truth*, *Public Liberty Radio*, and *The Journey*, it's very obvious that by mocking radio talk show hosts like Glenn Beck and Sean Hannity, they are making a commentary about how males have come to idolize this hyper-masculine model of what a man should be: [55] one host talks about the Mythic female orgasm and how it doesn't exist. He goes on to talk about how his wife took up tennis lessons because he doesn't satisfy her, and she works out with her trainer two or three times a day. They CLEARLY are mocking the masculinity archetype of being so masculine that they don't even believe females can have orgasms because they've never tried. [56]

This biting satire is funny and makes the player feel intelligent and savvy in the same way *The Daily Show* and the *Colbert Report* reward the viewer who can understand these shows' analyses of world events. The player plays at hyper-masculinity while the environment, itself, is mocking it. [57]

CONCLUSION

Given the massive shifts in the gender cultural landscape, it is to be expected that there is some confusion about what kinds of behavior are to be expected, desired, and rewarded in males. This can largely be seen as a good thing. It means that gender norms are changing and there is much more flexibility for males to behave in non-traditionally masculine ways. Nonetheless, nostalgia for the traditional male persists, as is evidenced by a trip to the movie theater, a glance at magazine racks, a few hours watching television or any of the commercials for the U.S. Marine Corps. Given the persistence of such nostalgic ideas of what it takes to be a man, *GTA* gives players the opportunity to "play" at being one. It allows players to feel that they "have a pair," and that testicles still mean power and that it is a desirable kind of power. *GTA* does so in a very different way than *Call of Duty* or other games that glorify and perpetuate such mythologies without questioning them. It enables the performance of hegemonic masculinity while simultaneously demonstrating how ridiculous and dangerous it is.[58]

NOTES

1. Robert M Sapolsky, *A Primate's Memoir* (New York: Scribner, 2001).
2. Desmond Morris, *The Naked Ape: A Zoologist's Study of the Human Animal* (New York: Dell, 1967).
3. Rockstar Games, *Grand Theft Auto IV* (Take-Two Interactive, 2008).
4. Joel Roberts, "Australia Bans 'Grand Theft Auto.'" *CBS News*, July, 29, 2005, accessed at http://www.cbsnews.com/stories/2005/07/29/tech/main712700.shtml.
5. United States Securities and Exchange Commission, "Form 8–K Take-Two Interactive Software, Inc. 0–29230" (Washington D.C.: SEC, 2008).
6. Eric Magnuson, *Changing Men, Transforming Culture: Inside the Men's Movement* (Boulder, Colo.: Paradigm, 2007).
7. Michael S. Kimmel, Jeff Hearn, and Raewyn Connell, *Handbook of Studies on Men & Masculinities* (Thousand Oaks, Calif.: Sage Publications, 2005).
8. Susan Faludi, *Stiffed: The Betrayal of the American Man* (New York: W. Morrow and Co., 1999).
9. Raewyn Connell, *Masculinities, 2nd ed.* (Cambridge: Polity Press, 2005): 55–56.
10. Connell, *Masculinities.*
11. Internet Movie Database, "Yearly Box Office: 2008 Worldwide Grosses" (Internet Movie Database, 2009), accessed at http://boxofficemojo.com/yearly/chart/?view2=worldwide& yr=2008&p=.htm.
12. Jon Favreau, "Iron Man" (Paramount Pictures, 2008).
13. Christian Riesen, Pistolhot, Zephyr1991, Jimmydanger, Donkarnage, et al. "Grand Theft Auto IV: Story Walkthrough." (gamewiki.net, 2008), accessed at http://gamewiki.net/Grand_ Theft_Auto_IV/Story_Walkthrough.
14. Brian Limond, "GTA IV Prostitutes: How Do You Kill Yours?" accessed at http:// www.youtube.com/watch?v=n7b9SbFzIp0.
15. Associated Press, "Clinton Seeks 'Grand Theft Auto' Probe." *USA Today*, July 14, 2005: 3.

16. Elena Bertozzi, "'I Am Shocked, Shocked!' Explorations of Taboos in Digital Gameplay." *Loading . . . The Canadian Journal of Game Studies* 1:3(2008), accessed at http://journals.sfu.ca/loading/index.php/loading/article/view/27/0.

17. Spencer Striker, personal communication, expert player review of paper (25 Apr. 2009).

18. Malcolm Vick, "Danger on Roads: Masculinity, the Car, and Safety." *Youth Studies Australia* 22, no. 1(2003): 32–37.

19. Center for Disease Control, "Wisquars Details of Leading Causes of Death" (edited by National Center for Injury Prevention and Control. Washington, D.C.: U.S. Government, 2006).

20. Bertozzi, "'I Am Shocked, Shocked!' Explorations of Taboos in Digital Gameplay."

21. Tom Evans, "The Cars of Grand Theft Auto 4" (Microsoft Network, UK), accessed at http://cars.uk.msn.com/News/car_news_article.aspx?cp-documentid=8334479.

22. Robert Worth, "Saudis Race All Night, Fueled by Boredom." *New York Times*, March 7, 2008, accessed at http://www.nytimes.com/2009/03/08/world/middleeast/08drift.html?scp=1& sq=arab%20drag%20racing&st=cse.

23. Diana Fuss, *Essentially Speaking: Feminism, Nature & Difference* (New York: Routledge, 1989).

24. Joel Berger, *The Better to Eat You With* (Chicago: University of Chicago Press, 2008).

25. Rong Jiang and Howard Goldblatt, *Wolf Totem* (New York: Penguin Press, 2008).

26. Roger Horowitz, *Boys and Their Toys?: Masculinity, Technology, and Class in America* (New York: Routledge, 2001).

27. Sporting Good Manufacturers Association, "Sports Participation in America." 2008, accessed at http://www.sgma.com/reports/228_Sports-Participation-in-America-2008.

28. Lisa A. Williams and David DeSteno, "Pride: Adaptive Social Emotion or Seventh Sin?" *Psychological Science* 20, no. 3 (2009): 284–88.

29. Williams and DeSteno, "Pride: Adaptive Social Emotion or Seventh Sin?"

30. Derek Campbell, "Grand Theft Auto IV Cool Tips and Tricks" (monkeysee.com, 2008), accessed at http://www.monkeysee.com/play/10158.

31. Noah J. Sandstrom, Jordy Kaufman, and Scott A. Huettel, "Males and Females Use Different Distal Cues in a Virtual Environment Navigation Task." *Cognitive Brain Research* 6, (1998): 351–60.

32. Georg Grön, Arthur P. Wunderlich, Manfred Spitzer, Reinhard Tomczak, and Matthias W. Riepe, "Brain Activation During Human Navigation: Gender-Different Neural Networks as Substrate of Performance." *Nature Neuroscience* 3 (2000): 404–8.

33. Many thanks to Cale Reid, Nick Nieuwenhuis, and Matt Lahl for their assistance in playing *GTA IV.*

34. Connell, *Masculinities.*

35. Apatow, filmography available at http://www.imdb.com/name/nm0031976.

36. Campbell, "Grand Theft Auto IV Cool Tips and Tricks."

37. Limond, "GTA IV Prostitutes: How Do You Kill Yours?"

38. Charlotte Watts and Cathy Zimmerman, "Violence against Women: Global Scope and Magnitude." *The Lancet* 359, no. 9313 (2002): 1232–37.

39. blip.tv, "Grand Theft Auto 4 Leaked Gameplay" 2008. Accessed at http://gameplayvideos.blip.tv/file/858225/.

40. Roger A. McCain, *Game Theory: A Non-Technical Introduction to the Analysis of Strategy* (Mason, Ohio: Thomson/South-Western, 2004):183.

41. Michael A. Messner, Michele Dunbar, and Darnelle Hunt. "The Televised Sports Manhood Formula." *Journal of Sport and Social Issues* 24, no. 4 (2000): 380.

42. Messner, Dunbar and Hunt, "The Televised Sports Manhood Formula," 392.

43. Suzanne E. Hatty, *Masculinities, Violence and Culture* (Sage Series on Violence against Women: Sage Publications, 2000); Fuss, *Essentially Speaking*; Elizabeth Grosz, *Volatile Bodies* (Bloomington: Indiana University Press, 1994).

44. Michael J. Clark, "Faludi, Fight Club, and Phallic Masculinity: Exploring the Emasculating Economics of Patriarchy." *Journal of Men's Studies* 11, no. 1 (2002): 65–76.

45. David Fincher, "Fight Club" (Art Linson Productions, 1999).

46. John Mello, "Haitian Group Files Suit to Ban 'Grand Theft Auto: Vice City'" *TechNewsWorld* (2004), accessed at http://www.ecommercetimes.com/story/32527.html.

47. Lisa Nakamura, *Cybertypes: Race, Ethnicity, and Identity on the Internet*. (New York: Routledge: 2002).

48. Anonymous, "GTA IV Leaked Gameplay #2" (blip.tv, 2008), accessed at http://blip.tv/file/858469.

49. Robert A. Nye, *Masculinity and Male Codes of Honor in Modern France* (Berkeley: University of California Press, 1998); Judith Kegan Gardiner, *Masculinity Studies & Feminist Theory: New Directions* (New York: Columbia University Press, 2002).

50. Wikicheats, "Grand Theft Auto IV—Pc Ps3 Xb360/Walkthrough" (Gametrailers.com, 2008), accessed at http://wikicheats.gametrailers.com/index.php/Grand_Theft_Auto_IV_-_PC_PS3_XB360/Walkthrough.

51. blargu, "Sniper Fun in GTA 4." (youtube.com, 2008), accessed at http://www.youtube.com/watch?v=wlWLs-AyvO8&feature=related; Green, Michael. Personal communication, expert player review of paper, email, May 9, 2009.

52. Johan Huizinga, *Homo Ludens: A Study of the Play-Element in Culture* (Boston: Beacon Press, 1955), 65.

53. Huizinga, 90.

54. Downtown Jimmy, "Grand Theft Auto Iv (Gtaiv) Pg 2/2 Xbox 360/ Ps3 Review" (Extreme Gamer, N.d), accessed at http://www.extremegamer.ca/multi/reviews/gtaivb.php.

55. pesiru93. "GTA IV—We Know the Truth(Wktt) Radio Broadcast. " (YouTube 2009), accessed at http://www.youtube.com/watch?v=aBUsTfbjVok.

56. Chris Bates, personal communication, expert player review of paper via email, May 9, 2009.

57. This is perfectly exemplified by the ad for Powerthirst that can be found on YouTube at http://www.youtube.com/watch?v=qRuNxHqwazs. A similar ad airs on TV in *GTA IV*.

58. Many thanks to Chris Bates, Michael Green, and Spencer Striker (expert GTA players) for their feedback and commentary on this paper.

Chapter Two

Discursive Engagements in *World of Warcraft*

A Semiotic Analysis of Player Relationships

Elizabeth Erkenbrack

When considering ethnographic and discursive research of online communities, it is essential to understand the structure of interrelationships between participants within this digitized space. In addition to clarifying the relationship between various frameworks within which World of *Warcraft* (*WoW*) players are situated and the social realities the players are co-constructing through their discourse, this paper will explore specific discursive dynamics of player relationships. Most particularly, I will be looking at how the relative positionality between players is both created and represented, the multiple commitments of players evidenced in their speech, and finally the discursive evidence calling into question the assumed discrete separation between the online and offline worlds.

WoW offers an illuminating example of the intricate social behavior and patterns at work in online communities. *WoW* is a Massively Multiplayer Online Role Playing Game (MMORPG) that boasts over 11 million players, a widely influential economy,[1] and a virtual world[2] entitled Azeroth, making it both the largest MMORPG ever created and an excellent nexus of social and discursive information. The average player spends 22.7 hours per week engaged in the game[3] through his or her avatar,[4] and most players spend several additional hours on the discussion boards and forums related to the game. Beyond these engagements, fan art, fan videos, and endless *WoW* blogs, wikis, and websites abound, all created and maintained by *WoW* players. These are often used simultaneously, with players checking wikis or blogs while playing the game, in addition to the constant out-of-game communication between players through Ventrillo or Skype. This simultaneous engagement within the game along with other frames of participation high-lights a player's ability to co-exist and orient to a variety of participation frameworks[5] throughout gameplay. Thus, the assumed distinction between in-game and out-of-game becomes complicated and the mutual effects on behavior between avatar and player also become clearer. The player and avatar are indexically linked while the player is signed into the game. Avatar movement is made possible through the player action, and the motivation behind the player keystrokes is often informed by online—not necessarily in-game—details such as blog entries or advice from discussion boards. For example, how a player decides to engage their avatar in battle is often based on advice from blog and wiki sources. In addition, a great deal of behavior is certainly motivated by in-game interaction and voice or instant-message dis-cussions between players. Attempting to negotiate these supposed boundar-ies reveals a nuanced understanding of relationships between *WoW* players and the game media, as well as the social work these discursive relationships are doing.

The focus of this work is to explore MMORPGs as a particular type of mediatized discursive activity with connective tacit social structures, using *WoW* player discourse as a specific example. Through applying traditional sociolinguistic approaches to this particular type of online technology, I am able to not only alter and clarify appropriate methodological approaches for mediatized interaction, but also tease out a greater understanding of the so-cial work that is occurring within these spaces. These digital cultural spaces reveal how new enactments of linguistic ideologies and social structures provide salient examples of the understandings (and misunderstandings) widely harbored regarding online mediated interactivity.

Precise language use in analysis of these virtual spaces is absolutely vital. The dichotomy between "real" and "virtual"[6] implies a discreteness to these frameworks of interaction that is methodologically misleading. It is also dangerous to refer to either online or offline environments as "worlds," virtu-al or otherwise, as each involve participation frameworks that are tacitly and explicitly linked to each other. This is seen through the fact that an avatar's movements and actions are only possible through the keystrokes and mouse-clicks of the player. This can also be seen in the linguistic marking of the performative role of *player* or *gamer* which links the person in a non-digital participation framework to the wider digital participation framework of the MMORPG. The real-time interaction of the avatars, made possible through the indexical connections of keystrokes and mouseclicks, thus enables a real-time interaction of the players themselves because the avatars and players are inextricably linked. For some specific research questions, especially those relating to the economics[7] or coding[8] aspects of these games, this distinction becomes more relevant. However, for questions of social and linguistic work in creating and maintaining personhood, community, identity, and communication, it creates a dichotomy that is not useful as a theoretical foundation.

Since language use in analysis of MMORPG phenomena is so important, for this paper I will refer to *WoW* as an MMORPG, not a virtual world, and make no attempt to linguistically distinguish between the online and offline environments, since they motivate and inform each other. I will entirely avoid making any distinction between "real" and "virtual"[9] or "synthetic"[10] environments with regard to these spaces, and argue that these descriptors hinder research of positionality and mutually oriented behaviors in digital environments. To help highlight interrelationships between these frameworks, my analysis will be based within the theoretical scope of semiotic registers and discursive footing.

REGISTERS, ENREGISTERMENT, AND DISCURSIVE FOOTING

For this chapter, I am interested in what social work discourse oriented to *WoW* is actually doing. To this end, I will be invoking Agha's[11] concept of registers as "reflexive models of language use that are disseminated along identifiable trajectories in social space through communicative processes."[12] As a reflexive model of language use, registers consist of both linguistic and non-linguistic signs that are associated with a particular social type of person. A rather simplistic, but effective, example can be seen in an individual dressed in scrubs with a stethoscope around his/her neck perhaps saying "50 CCs" or "endocardiogram" or "it seems to be acute sinusitis." The linguistic signs work with the non-linguistic signs to position this particular individual as a doctor or other medical professional. Many individuals can recognize this without needing to be told the person's profession because of our social ability to recognize embodiment of registers. Register analysis allows us to see the dynamic process of language through time and among users since language is itself a particular model of personhood. Thus, encounters with registers are "encounters with characterological figures stereotypically linked to speech repertoires (and associated signs) by a population of users."[13] Importantly, registers illuminate performed personhood far beyond professional positions like the doctor. Often the way one speaks will cause others to make assumptions about, perhaps, education level, area of origin, and economic status, just to name a few. In other words, the reflexive models of language use—which I understand registers to be—serve to dynamically create and perpetuate structures of personhood within particular populations. As I will show, the transcribed discourse presented in this paper is not only indicative of the social positions that the individuals inhabit within the situation, but it also reveals the consistent negotiation of this footing.[14] The muted materiality of this discursive process becomes apparent through the lens of semiotic register analysis. Through this analysis, I am able to see various aspects of how the discourse is revelatory of the emergent social relationships that are constructed through language use.

MMORPGs are necessarily multi-faceted discursive engagements since they involve multiple social environments. Registers are a way of beginning to understand what the language use is actually doing in terms of social reality. In *WoW*, the many different layers of discourse include, but are not limited to the "WoW register." The resultant text, as represented in this paper through the transcript, has both denotational and interactional coherence.[15] That is, in order for this communication to have been successful,[16] the individuals need to actually understand what is being said (denotational coherence of discursive reference) and to recognize their socially organized regularity (interactional coherence) within the exchange. Within the game, individuals need to understand particular vocabulary and grammar unique to *WoW* as well as the relationships to each other as players and avatars. These various elements are all apparent within the excerpts of text presented later, as they are in any successful socially discursive exchange. In addition to the necessary coherence, the use of voice[17] indexes particular types of personae. Speech forms are then semiotic expressions that can materially link the speaker to typifiable images of personhood. Not only is successful discourse denotationally and interactionally coherent, but it is also doing social work among the participants by linking speakers to particular inhabitable images of personhood through the voicing of registers.

Understanding voicing allows me to look at the process of role alignment, or "patterns of relative behavior . . . focus[ing] on the expression of voices and figures in the behaviors."[18] Footing, then, is a particular type of role alignment that focuses on participation in "coordinated task activity."[19] *WoW* play is a fascinating example of this since the technology enables such an activity in real time without the necessity of face-to-face interaction. Role alignment, evidenced through the voicing, which is itself implicitly enregistered, allows me to make more transparent the social negotiations at work through interactive use of discourse. Both Hanks[20] and Irvine[21] also show that the complexity of role distinctions within a particular discursive context are highly varied and intrinsically connected to the communicative nuances in the event. To be able to analyze discourse then necessitates access to the cues within the event understandable only through contextualization. As a result, studying discourse necessitates an understanding of the context in which the language is being used and the social work that is being done.

A MMORPG is an unusually interactive social media. Ethnographic and discursive analysis of this type of media necessitates an understanding of the technology and its effects to ensure that it is recognized as both a cultural artifact as well as an actual site of culture and cultural formation[22] with intertextual semiotic structures.[23] Rather than treating the internet as simply an example of material culture[24] or as a site of ethnographic research in and of itself, excluding analysis of the players[25] behind the avatars,[26] I will use discourse analysis to argue that understanding the social work occurring within these technologies necessitates a particularly nuanced understanding of the interactive realities at play.

DISCOURSE CONTEXT

In order to provide a "plausible, minimally rich account of the meaning and effectiveness of discursive interaction that has some predictive power"[27] the context to a discursive event must be presented. In the particular excerpts that I discuss next, the transcription covers a group of *WoW* players as they prepare and implement an attack on a boss in *WoW*.[28] There are a total of 12 speech participants, although not all are represented in these excerpts, and importantly not all geographically co-present. In the room with me, the re-searcher, were 5 individuals: four *WoW* players on their computers—repre-sented as K, M, Jo and S—and a friend who plays *WoW* but is not playing in this instance, represented as R. All five have logged hundreds or thousands of hours on the game and have fought this particular boss[29] multiple times. I (L) am videotaping, audio taping, observing the play and participating in the conversation as a researcher and a less experienced player of *WoW*. In addi-tion to the 6 of us in the room, there are 6 additional players—represented as A, Cr, J, Je, Ma, and C—who are not in the room but whose avatars are co-participating and whose voices are heard over Teamspeak.[30] While these six are geographically situated in Florida, New York, and Paris, they are audibly present through Teamspeak and visually co-present through their avatars. The 10 players are able to speak to each other through Teamspeak and type to each other in the game in instant-message fashion. I captured recordings of all the discourse both in the room and over Teamspeak as well as video-captured the discourse on-screen. Both myself and R, the other non-player, were only able to participate in the in-room discourse. Thus, the transcript reflects three different levels of communication: the in-room (but not Team-speak) linguistic and non-linguistic communication, the Teamspeak dis-course, and the in-game typed discourse. Adding to the complication of this already layered exchange is the fact that the game itself produces some of the typed, im-esque discourse in the game, voiced by the computerized charac-ters the players encounter.

To summarize, within the transcript there are four laminated types of discourse within this particular event. In the continued interest of clear language choices in my analysis, I will be using a. on-screen (OS), b. in-room (IR), and c. Teamspeak (TS) as descriptors of these different frames of discourse. Recall that these are only to clarify different frameworks of social participation and not meant to devalue the interrelatedness and blurred lines among these frames. For example, discourse by the four players in the room can simultaneously be Teamspeak and in-room, the difference being the size of the intended audience. The transcript itself indicates three different frames for the four types of discourse, collapsing the on-screen discourse of the players and the computerized characters since they appear the same within the game and inform behavior in similar ways. With this context in mind, we can begin to explore how interpersonal relationships are negotiated, polycentric commitments of players are made apparent, and the influence that the discourse has on behavior is various frames.

WHAT DOES THE DISCOURSE ILLUMINATE AND REINFORCE?

Interpersonal Alignment

In the excerpt that follows, I show both the ongoing negotiation among players within the game and the presupposed knowledge evident in word choice and assumed behavior. The enregistered voices,[31] or voices tied to a register, link social character to an individual's discourse. As with every social interaction, the players present selves that are figures constructed and projected through narratives oriented to past encounters.[32] This relationship between the self and the narratives informs interactions between the players and formulate the interpersonal footing at work. The dynamically constructed relationships are then inherently linked to previous encounters. Previous experience within *WoW* provides an understanding of the terminology, requisite behavior, and necessary group dynamics for all the participants. In this example we will look at emergent forms of expert and novice positionality. For example, in *WoW*, expertise is based on a combination of time commitment, previous experience in any given encounter, and the ability to appropriately utilize an avatar's abilities. In any given situation where multiple individuals are playing together trying to beat a boss, the roles of expert and novice will come into clear relief. In this excerpt, while all of the participants have logged hundreds of hours in the game, A has never fought this particular boss before. Jo and M, in contrast, have both played against this boss many times and also play avatars which are well known within the game as being two of the most elite. As will be evident, Jo and M are both oriented to the preparation of A for the upcoming fight as well as constantly negotiating their mutual expertise.

JoTS: Other thing! Elementals. He will summon a bunch of water elementals.

SIR: Assuming we get that far.

JoIR: Right, assuming we get that far [*gestures toward S*].

MIR: Shh! Positive thinking [*points at S*].

JoTS: When he does that, you need to banish one and fear one. And that's the game.

ATS: I'll get by.

Within this brief excerpt, presuppositions and entailments[33] are at work in order to achieve interactional coherence. In other words, certain types of knowledge are assumed to be present and other knowledge that will be necessary to beat the boss later is presented. The presupposed knowledge of the game is found in the unexplained understanding of what an elemental is, especially a water elemental, as well as what the actions of banish and fear mean. This is interesting because it presupposes A's familiarity with the game while also acknowledging her unfamiliarity with the particular situation and how her avatar should fight these elementals. The entire group is focused on the interactional coherence both in the explanation process—indicated through their silence over Teamspeak throughout the explanatory process—as well as the recognition of this explanation's vital role in successful coherence once the fight starts. At this time, M's expertise does not confront Jo's expertise in the explanatory process, but he does begin to implicitly contribute his own voice and claim positionality through his imperative statement.

The in-room speech is particularly interesting from two perspectives: the first being that the three participants are all also players so they are signed into Teamspeak but choose specifically to keep their exchange in-room. While S is articulating a doubt in the group's potential success, and perhaps does not want to share the concern with the entire group, particularly A, there is an alternate explanation based on social cohesion. This being that S does not want to confront Jo's domination of the Teamspeak channel. No other players have been on Teamspeak aside from Jo at this time, behavior that is based upon presupposed knowledge of pre-fight coordination within the game. When a group is preparing to attack a boss, the leader(s) or expert(s) within the group dominate the voicechat channel in order to quickly coordinate everyone. If someone at this point has a comment or concern, they are expected to keep it off of the voice chat and communicate it either by IM or in-room communication. While this rule is quite logistical—having 10 voices speaking at once would be cacophony—it also has social structuring implications regarding the relative positionality of players based on their access to this voicechat. Importantly, S does not occupy a purely acquiescent role. In fact, she directly challenges Jo's assumption of the group's success and Jo himself acknowledges the validity of her concern both linguistically and gesturally. Thus S is simultaneously maintaining the wider social assumptions of access to voicechat during preparation time, but still negotiating her own positionality by revealing a nuanced understanding of the game and the possibilities for why the group might not be successful. M, another widely recognized expert within the game and the group, challenges both S's comment and Jo's acknowledgement. His deictic gesture of pointing seems to indicate that S is the addressee of his order to have positive thinking, but Jo is implicitly addressed as well through Jo's own alignment with S regarding the pessimism of reaching the elementals. However, M refrains from a public challenge to either S or Jo, despite his expert positionality and resultant ability to access Teamspeak. At this point, his use of humor is not meant to be an order from an expert as much as it is encouragement and emotional preparation against S's concern. Once Jo continues with the explanation over Teamspeak, A, the addressee of the Teamspeak instruction, simply acknowledges recognition of herself as both addressee and as novice as well as providing reassurance that she understands the instructions and her role.

What this shows is the understood relationships between the players are continually negotiated in the various frames and are built on understandings of role alignments and interactive behavior appropriate within the context. Displays of assumed knowledge, goals of the discourse, and audience size are all informative to the models of personhood that they inhabit and the resultant interactive positioning. As Agha articulates regarding the interactive implications of register use, "any use of a register performatively models specific footings and relations between speaker and coparticipants."[34] In understanding this process of relationship development between individuals through their various uses of language, it is essential to remember that role alignment occurs not between the persons themselves but between the figures performed through the speech. Jo's presence in Teamspeak and skilled deployment of the *WoW* register model his role as expert, while A's acknowledgement marks her relative role as novice. S models an experienced player in her understanding of the game and the group dynamics, and her assumption of failure. M is not modeling ignorance in his confrontation to her pessimism, but rather understands his *WoW* reputation as an expert and is able to use it to challenge and encourage S's pessimistic opinion and Jo's subsequent alignment with S. The voices and behavior displayed created models, or figures, through which the individuals align. And these behaviors, discursive or otherwise, are contextually based. So, while an individual might be discursively and socially constructed as a novice in this discourse, it is the alignment between the role of novice and the expert, rather than the alignment between the individuals inhabiting these roles, that are of social significance. The group is oriented to the task at hand, so Jo's modeling of expert and A's novice positionality are based within this context, and doing significant social work therein.

Polycentricity

Relationships between models of personhood are not the only commitments that a speaker maintains throughout a discursive exchange. For example, Blommaert's[35] concept of polycentricity in which various patterns in speech are indexically linked to different frameworks is also evident in this discourse. While his argument focused especially on multilingualism and the various patterns of authority that multilingual individuals orient to, his concept of various personal commitments throughout a communicative process remains informative in this instance. For example, as shown by the excerpt below, the players remain oriented to both on-screen behavior and to in-room behaviors and interactions.

RIR: Do we have no melee DPS?

STS: Nope.

MTS: Shh! We have kitties [*looks at R*].

CTS: We have some fearing and Lisara Vjing.

RIR : Like, melee DPS is what destroys them.

MaTS: Could someone . . . it sounds like there's a steamer engine going on there in the background [*general laughs*].

MIR: Do do do do do [*pumps fists*].

JoTS: I had to steal wireless from the train station because here in my apartment.

MIR: Shh! Shh! Shh! Pump it! [*All listen.*]

KIR: My new garden [*in time with music, glances at camera*].

JoTS: Are we good to go?

Throughout the excerpt, the players are oriented to the game, to each other, and to me. Within the different frames, there are different levels of authority among the players that are clearly being negotiated. R, the *WoW* player who is not playing at the moment, reinvigorates S's prior concern over whether or not the party will be successful, asking about melee DPS. Evident throughout, the assumption of understanding the *WoW* register is assumed. Not only is it assumed that the group knows what melee DPS is (it is a type of inflicted damage), but the implications of not having melee DPS are understood. After S confirms the lack of melee DPS, M responds to both over Teamspeak, continuing his ongoing quest of encouragement, shushing again and reassuring that there are kitties. M clearly positions himself as both a positive reassurance within the context and an expert, offering the solution of a hunter's pet in response to a lack of a different type damage. He is so confident in his response that he expands the conversation to include the non-present participants over Teamspeak. C at this point joins, offering additional options for how the group can compensate for a lack of melee DPS. S does not respond, but R does, despite his lack of in-game involvement, reasserting that melee DPS is what is needed to win. As the interpersonal alignments are being negotiated regarding who best understands what is needed to win, the tacit polycentric orientations of the players come to the forefront, made most clear when Ma asks for the music to be turned down.

Ma's clear reference to an out-of-game variable that is affecting the communication and, as a result, the interaction of the team, leads to a particularly salient stretch of discourse. After everyone laughs at the joke within which the request is embedded, M starts echoing the music even while it gets turned down and even tries to silence Jo's explanation of the noise. K joins M's echoing of the music, acting on in-room interactive orientation as well as physically recognizing the camera through her shift in eye-gaze. Jo, however, does not engage in the music, reorienting the discourse and regaining his position as both expert and leader when he asks if everyone is good to go. M entirely avoided the response to R's continued concern regarding melee DPS and it does not get re-addressed, which is reflective of R's non-involvement in the immediate game, while not negating his knowledge of the game and situation. Throughout the conversation, both M and Jo navigate between their orientation to the game and the group (either in arguing for substitutes to melee DPS or to re-orient the entire group to the online attack), but also to their orientation as residents of the apartment in which the 6 co-present individuals are situated. The players clearly switch their use of discourse as they shift their social focus, such as discontinuing use of the *WoW* register when not discussing the game. These particular points of discourse act as data of social life, which themselves "point to lived moments that lie beyond them."[36] In order for the meaning in this discourse to remain coherent, under-standing the context and the multiple frames is important, but equally neces-sary is the understanding that there are multiple commitments on the part of the speakers to these various frames. Behavior on-screen is not always moti-vated by the same interactional goals as behavior between players not orient-ed to in-game processes. But, importantly, players can be simultaneously committed to both of these goals and switch between them as we see.

These players are oriented to polycentric centers of interaction based on the different frameworks of interaction. Since language is linked to the cultu-ral models and behavioral patterns, use of multiple discourses allows us to see the multiple positional commitments on the part of the speakers. The shifts in and out of the *WoW* register as well as the implicit interactive goals of the discourse illuminate how these *WoW* players are themselves shifting in their social commitments. The laminated frameworks enable shifting be-tween them from one turn to the next, so that the coherence of communica-tion doesn't falter despite the implicit shifts in social goals.

On-Screen and In-Room Discursive Effects

Clearly, while constructing MMORPGs as discrete "worlds," separate from the out-of-game environment lays theoretical and foundational stumbling blocks, they are two participation frameworks with different social goals on the part of their participants. In looking at the two frameworks, we find that they mutually affect both discursive and behavioral patterns. Discourse on screen will affect in-room behavior, and vice versa, as evidenced below. This situational analysis shows how different discursive frames can interact with each other, as well as reiterating the mutual influence between the on-screen and in-room frames. This short excerpt illuminates how the computerized discourse of the in-game character has influences on in-room discourse and behavior which then manifests into on-screen behavior as well.

Shade of Aran OS : I'll show you this beaten dog still has some teeth!

JoTS: Flame wreath.

MTS: Flame wreath, oh God!

JoTS: Flame wreath, nobody move.

MTS: Do not move. Do. Not. Move.

[All avatars stop moving.]

At this point in the discourse, the fight between the 10 characters being controlled by the players and the computer-generated boss named Shade of Aran has definitely begun. By looking at the present shifters, [37] as well as the behavioral influences, one can see how the frames of in-game and in-room are separate but mutually informative. Shifters are lexical items that change referential meaning entirely depending on the situation, an example being the pronoun "I" which changes its meaning every time a speaker changes. The use of shifters in this example manages to interestingly collapse the two participation frames. The implicit 2nd person object of the imperative sentences, in this case the "you all" who are being told not to move, supposedly refers to the characters but implicitly refers to the players who control the characters' movements. Thus, the "do not move" followed by the stilling of the avatars indexes the simultaneous stilling of the player's left hand so that the avatar doesn't move. The shifting object of this imperative statement in fact equates the avatar and player behavior, highlighting the indexical link between the two.

Further, the behavioral influences between on-screen and in-room discourse further help us see that the two frames are quite mutually influential on behavior of both the players and avatars. This is most clearly seen in the reaction to Shade of Aran's yell. This statement has two immediate effects. On screen, a series of flame wreaths appear around many of the characters on screen. Interpersonally, Jo immediately recognizes the in-game implications of the Shade's statement, articulates what that behavior will look like and further instructs the other players should all play their avatars as a result. Each of Jo's three reactions is a behavioral reaction to the in-game discourse. In other words, while a "virtual world" and the "real world" are frequently framed as discrete areas or locations, for social and discursive work they are quite simply two interwoven frames of participation without any unique discreteness.

Within this cross-frame discourse, there are many levels of discursive expertise embodied. While every English speaker likely understands the meaning of the statement "do not move," the players must also understand that this means that their character should be stilled. However, they also need to understand that they are expected to continue playing the game in other ways (e.g., characters shooting spells, healing, etc.). The denotational congruence of understanding both what a flame wreath is, and what "do not move" means in terms of a character's behavior are both implicitly understood by users of the *WoW* register. Further, we return to the interactional congruence of every character stopping in place yet continuing to heal and attack while unmoving, thus preventing the group from dying. These linguistic and non-linguistic behaviors are all interdiscursively linked, [38] creating continuity between the frameworks and drawing on understanding this type of discourse through presupposition of prior experience. Understanding how to navigate the multiple frameworks necessitates previous experience in the navigation process. Players begin learning to do this in the game gradually, starting with two avatars playing together to beat a boss they can't beat individually, followed by five avatars bosses, then fighting bosses that require ten individuals. This cross-framework communication is progressively learned through time and experience. The very ability of the players to easily orient to the multiple frames, and orient themselves to various goals among these frames, itself marks their expertise in the game. As shown throughout the various excerpts presented, the high level players are able to align to different frameworks, shift these alignments quickly, maintain understandable communication, and navigate the cross-frame discursive influences.

While each of these examples is evocative of interactive work, polycentric orientations or multiple frameworks informing behavior, all three of these types of social work co-occur within every instance of discourse, as made very clear in our final excerpt.

Intererpersonal, Polycentric, and On-Screen/In-Rom

JoTS: Blizzard [*pause*].

Shade of Aran OS: I'll freeze you all!

JoTS: Water elementals are up! Banish, fear, etc. Everyone . . .

Shade of AranOS: I'm not finished yet. No, I have a few more tricks up my sleeve . . .

MTS: Just pick whichever one you want and banish it.

KIR: Are we zerting them, or . . .

JoIR: You've got enough fears to.

KIR: I used my two, but . . .

RIR: This music does this flame scene amazing.

[*JolladOS has died.*]

MIR: It's hilarious.

Shade of AranOS: Torment me no more!

KTS: Down.

Jo IR: Is that fire? Is he doing fire?

MIR: Arcane missiles.

This last excerpt of discourse helps to illustrate how the three aspects of discursive behavior that I examined—the dynamic interpersonal relationships, the polycentricity and the on-screen/in-room informativity—are all co-present. The more complex the interaction, the more the skill with which all of these participants navigate the laminated frame and multiple social goals is shown.

I will begin with an examination of some of the shifters, again looking at an imperative statement. When M states "Just pick whichever one you want and banish it," the understood addressee seems to be the character that A is playing. This is clear to everyone present because A is playing the only Warlock,[39] thus her character is the only one that can banish creatures. Thus, it cannot apply to anyone else, which everyone understands. Interestingly, however, the referent of *you* changes within this single sentence. The "you want" clearly references the player A, since a character cannot actually want. However, the implied you in "[you] banish" clearly references the character since people cannot banish. Within this single utterance, looking at the tacit addressee reveals the close lamination of these different social frames. The player A will decide what she wants to attack and her implicit indexical link to her character negates the discursive necessity of singling the actor for the resultant banishing. This close discursive link also highlights the importance of research including both player and avatar in order to illuminate the nuanced social work being done.

Work of interpersonal expertise is also being done quite explicitly here. Throughout the conversation and this analysis, Jo and M have been navigating their dual expertise within the group and within the general game of *WoW*. Here, they are the only two to speak on Teamspeak throughout the attack until K's avatar dies and she announces it. Both Jo and M announce upcoming attacks and how to fight them to the rest of the group, with Jo announcing the upcoming attack first and M reiterating. However, this dynamic changes toward the end of the excerpt when Jo knows that the Shade of Aran's speech is warning him a specific attack is coming, but Jo cannot remember which specific attack it will be. He presents his guess to M, who corrects him, and then Jo proceeds to inform the rest of the group (discourse not included in this excerpt). Throughout, Jo and M continue to act in different frames as well, from M validating the music choice for the scene to Jo answering K's question as to how she can attack. The structures of authority shift between frameworks, and shift throughout the exchange, but the enregistered voices of Jo and M serve to position them as experts throughout this exchange.

The polycentric commitments of the players are on prominent display in this excerpt, especially in M's discourse. This is clearly seen in the seeming insensitive or nonsensical reaction of M when Jollad dies. Jollad is the avatar played by K, and M's immediate statement after the on-screen announcement that Jollad was killed is "it's hilarious." We have already seen that the players are continually aware of the on-screen discourse in addition to the verbal discourse, since it directly informs their behavior. Thus M is surely aware of the announcement of Jollad's death. Since he has been encouraging throughout, and explicitly trying to get others to think positively, this statement directly contradicts his previous discursive modeling, unless he is currently focused on an interactional goal in a completely different frame. Indeed, we can see that his statement is more likely responding to R's statement regarding the music playing as another fire wreath attack is about to occur. Although M's earlier statement in this excerpt directs in-game behavior over Teamspeak in order to achieve success in the attack, he shifts his center of focus to the in-room discourse in his ongoing attempts at encouragement to ease the interactive process. Over Teamspeak, he is committed to helping everyone act appropriately to beat the boss. In the room, he is including R in the interactive process and continuing the positive reinforcement and encouragement. Through understanding the polycentric orientations of his discourse we see that the model of personhood he inhabits throughout remains consistent, his discourse is simply oriented to interaction across multiple frames.

Throughout these various excerpts of discourse from a group of individuals oriented to *WoW*, it becomes possible to see the provocative and interesting work that can be done in these digital realms. It is apparent that participation in the higher levels of the game requires the ability to at least understand the *WoW* register. In addition, individuals are able to position themselves according to their presupposed and recognized abilities in conjunction with their effective and consistent use of this register. These positions, while structured, are also constantly negotiated. This vibrant locale of interactive possibility pulls into sharp clarity how these positionalities can be enacted even at the intersection of different types of communication. In looking at the mutual influence between various frames of participation at work, analyzing the discourse allows us to see how the various participation frames "leak"[40] into each other. Not only can the various commitments of the players be understood, but it is possible to also see how they move between these frames of communication. The interaction here is important for seeing how the on-screen and in-room communication affect each other, but arguably more intriguing is the theoretical implications of this understanding. The clarification afforded through approaching MMORPGs and other online games as sites of important social interactivity, but not discretely separable "worlds," is made abundantly clear here. To label them incorrectly imposes a predetermined restriction on the social and discursive possibilities of these spaces.

CONCLUSION

Throughout this chapter, I have explored evidence of the interactive realities, the multiple orientations of players, and the inter-frame effects of the online MMORPG *World of Warcraft*. In doing so, I have demonstrated how role alignment is manifested and managed, how successful interaction is achieved, and how discursive coherence is made possible even when communication occurs through various types of technology. This technology enables a coalescence of co-present specialized behavior, internet voicechat, and typed messaging. Moreover, these are all tied to register use, providing a dynamic tool through which to model oneself in relationship to others. In order to carefully dissect these questions, the implicit link between avatar and player—and the mutual informativity of this relationship—must be considered. To ignore either player or avatar is to miss much of the social and discursive information at work. As within any social setting, the commitments of the speakers shift easily and often without explicit comment, and any interactional (non)congruence can be attributed both to the footing and dynamic understanding of all present participants. As scholars continue to expand this exciting field and the interactive possibilities that the internet affords, we begin to be able to see how the discursive engagement of participants reveals the dynamic social work made possible through these technologies.

NOTES

1. Edward Castronova, "Virtual Worlds: A First-Hand Account of Market and Society on the Cyberian Front." In *The Game Design Reader*, ed. by K. Salen and E. Zimmerman (Cambridge, MA: MIT Press, 2001/2006), 814–865.

2. Virtual worlds are often defined as computer-simulated interactive spaces that present perceptual stimuli to the user, who in turn can manipulate elements of the modeled world. They often have "rules" like gravity.

3. Nick Yee, "The Daedulus Project" (nickyee.com, 2006), accessed at http://www.nickyee.com/daedalus/.

4. The term avatar to mean animated, three dimensional virtual bodies in virtual spaces was popularized by Neal Stephenson in his novel *Snow Crash* (1992). They are unique alter-egos through which a player can interact with the online environment.

5. Erving Goffman, "Footing." In *Forms of Talk*, ed. by Erving Goffman (Philadelphia: University of Pennsylvania Press, 1981), 124–159.

6. For a few previous examples of this dichotomy in use, see Tim Jordan, *Cyberpower: The Culture and Politics of Cyberspace and the Internet* (New York: Routledge, 1999); Howard Rheingold, *The Virtual Community: Homesteading on the Electronic Frontier* (Cambridge, MA: MIT Press, 2000); Sherry Turkle, *The Second Self: Computers and the Human Spirit* (Cambridge, MA: MIT Press, 2005); and Sherry Turkle, *Life on the Screen: Identity in the Age of the Internet* (New York: Simon & Schuster, 1995).

7. See Julian Dibbell, *Play Money: Or, How I Quit My Day Job and Made Millions Trading Virtual Loot* (New York: Basic Books, 2006); and Castronova, "Virtual Worlds: A First-Hand Account of Market and Society on the Cyberian Front."

8. Lawrence Lessig, *Code and Other Laws of Cyberspace* (New York: Basic Books, 1999).

9. See Richard Bartle, "Hearts, Clubs, Diamonds, Spades: Players who suite MUDs." In *The Game Design Reader*, ed. by K. Salen and E. Zimmerman (Cambridge, MA: MIT Press, 1996/2006), 754–787; and Julian Dibbell, *My Tiny Life: Crime and Passion in a Virtual World* (New York: Owl Books, 1998).

10. Edward Castronova, *Synthetic Worlds: The Business and Culture of Online Games* (Chicago: University of Chicago Press, 2005).

11. Asif Agha's use of register, as found in *Language and Social Relations* (Cambridge: Cambridge University Press, 2005); "Voice, Footing, Enregisterment," *Journal of Linguistic Anthropology 15*, 1 (2005): 38–59; and "Honorific Registers." In *Culture, Interaction and Language*, ed. by Kuniyoshi Kataoka and Sachiko Ide (Tokyo: Hituzisyobo, 2002), 21–63. See also Alessandro Duranti, *Key Terms in Language and Culture* (Oxford: Blackwell, 2001).

12. Agha, "Voice, Footing, Enregisterment," 38.

13. Agha, "Voice, Footing, Enregisterment," 45.

14. Goffman, "Footing."

15. Michael Silverstein, "The Improvisational Performance of Culture in Realtime Discursive Practice." In *Creativity in Performance*, ed. by K. Sawyer (Greenwich, CT: Ablex Publishing Corp., 1998), 265–312.

16. For discussions on successful and correct reference, see Keith Donnellan, "Reference and Definite Descriptions," *The Philosophical Review 77* (1966): 281–304.

17. Mikhail Bahktin, "Discourse in the Novel." In *The Dialogic Imagination: Four Essays,* trans. by Michael Holquist and Caryl Emmerson. (Austin: University of Texas Press, 1981), 259–422; Mikhail Bahktin, *Problems of Dostoevsky's Poetics,* ed. and trans. by Caryl Emerson (Minneapolis: University of Minnesota Press, 1984).

18. Agha, "Voice, Footing, Enregisterment," 53.

19. Goffman, "Footing," 144.

20. William Hanks, "Exorcism and the Description of Participant Roles." In *Natural Histories of Discourse,* ed. by M. Silverstein and G. Urban (Chicago: University of Chicago Press, 1996), 160–220.

21. Judith Irvine, "Shadow Conversations." In *Natural Histories of Discourse,* ed. by M. Silverstein and G. Urban (Chicago: University of Chicago Press, 1996), 131–159.

22. Christine Hine, *Virtual Ethnography* (New York: Sage, 2000).

23. John Paul Gee, *What Video Games Have to Teach Us about Learning and Literacy* (New York: Palgrave Macmillan, 2003/2007).

24. Daniel Miller and Don Slater, *The Internet: An Ethnographic Approach* (Oxford: Berg, 2001).

25. Tom Boellstorf, *Coming of Age in Second Life: An Anthropologist Explores the Virtually Human* (Princeton, N.J.: Princeton University Press, 2008).

26. Animated, three-dimensional virtual bodies in virtual spaces; unique alter-egos through which a player is able to interact with the online environment.

27. Silverstein, "The Improvisational Performance of Culture in Realtime Discursive Practice," 272.

28. In all transcripts, initials of players and names of avatars have been altered. The names of computer-generated characters (e.g., Shade of Aran) have not. As a reminder, in-room is notated as IR, on-screen as OS, and Teamspeak as TS*.

29. A boss is a character that is particularly difficult to beat, appears only once in the game, requires teamwork and a unique set of tactics to beat, and often marks progression throughout the game. Variation in the players' previous experience with this boss means some understand the tactics and teamwork necessary to beat him and others must learn their role.

30. Teamspeak is a "scalable" software that enables people to speak over the internet. Similar to a conference call, one must also have the correct sign-in to a server and channel information in order to gain access to the discussion. The technology is designed especially for gamers of MMORPGs.

31. Agha, "Voice, Footing, Enregisterment."

32. John Haviland, "'Whorish Old Man' and 'One (Animal) Gentleman': The Intertextual Construction of Enemies and Selves," *Journal of Linguistic Anthropology 15*, 1(2005): 81–94.

33. Silverstein, "The Improvisational Performance of Culture in Realtime Discursive Practice."

34. Asif Agha, "Introduction: Semiosis Across Encounters," *Journal of Linguistic Anthropology, 15*, 1 (2005): 1–5.

35. Jan Blommart, "Sociolinguistics and Discourse Analysis: Orders of Indexicality and Polycentricity" *Journal of Multicultural Discourses 2*, 2 (2007): 115–130.

36. Agha, "Introduction: Semiosis Across Encounters," 1.

37. Michael Silverstein, "Shifters, Linguistic Categories, and Cultural Description." In *Meaning in Anthropology*, ed. by K. Basso and H.A. Selby (Albuquerque: University of New Mexico Press, 1976), 11–56.

38. Michael Silverstein, "Axes of Evals: Token versus Type Interdiscursivity." In *Journal of Linguistic Anthropology 15*, 1 (2005): 6–22.

39. A Warlock is one of the ten classes of characters that a player can select for their avatar. Warlocks are members of the caster classes (so they are most powerful casting spells rather than hand-to-hand combat) which have powerful magical abilities including controlling an enemy's behavior, causing significant damage, or even healing. In this group, the Warlock was the only character able to banish the enemy.

40. Irvine, "Shadow Conversations."

Chapter Three

The Intermediate Ego

The Location of the Mind at Play

Vanessa Long

In this chapter, I illustrate a means of understanding the relationship between the user and video game which makes use of psychoanalytic and other contemporary theories. I introduce a concept that shall herein be known as the "intermediate ego" and demonstrate how it is created, maintained, and its importance to the gaming situation. I also demonstrate how it is utilized depending on the type of game being played by the user, and the type of effect this creates for said user.

Psychoanalytic theory has long been utilized to understand aspects of the world, but rarely has it been seen as a method by which to analyze video games, or indeed the virtual world in general. Despite this, theorists such as Rehak and Jagodzinski have paved the way into providing some understanding of games using psychoanalytic theory.[1] Furthermore, later theorists within the psychoanalytic field have applied their theories to a multitude of contemporary issues, such as Žižek's take on violence.[2] There has, however, been little psychoanalytic research into the virtual world. This appears unusual given how well psychoanalytic theory has been utilized in other aforementioned issues of contention.

The sense that few theorists have utilized psychoanalytic theory for this pursuit is difficult to comprehend considering that, according to recent reports, 53 percent of adults and 97 percent of teenagers in America play video games.[3] This alone makes it an important subject of debate in the current climate. Barthes[4] states that, "language can only obliterate the concept if it hides it, or unmask it if it formulates it." This is one of the reasons why psychoanalysis is ideal for analyzing games. Language is situated at the center of our foundations: our society, our ability to communicate effectively, our ability to read everyday objects in a correct way. However, from a psychoanalytic perspective, language can destroy in the same sense as it creates meaning. Items are read in similar ways based on our experience of those objects and things related to them. Using psychoanalytic theory has distinct advantages with relation to video games theory. In applying psycho-analytic concepts to what is occurring, we are able to perceive the structures at work for the duration of play. Furthermore, we might also perceive related elements that have an effect on us, even when we are no longer indulging in the act of play. I aim here to expose these structures, explain how it is that they come about during play, and how they are maintained throughout the play duration.

I draw upon past psychoanalytic works including those involving play spaces to show how these structures might come about and be maintained. Constructs such as these assist in our understanding of events that we would not otherwise necessarily be able to understand. Rather than attempting to analyze aspects from the surface, psychoanalytic constructs show a structure, one which we can build upon in order to gain new understanding. One must always remember, when dealing with psychoanalytic constructs and spaces, that they are not visible or identifiable in the same way as other constructs might be. While this makes them no less valid, it can make them more difficult to describe.

IDENTIFYING THE EGO

In psychoanalytic theory, there are three concepts that we must be familiar with in order to proceed: the id, ego, and the superego. I previously stated the difficulty in ascribing any physical nature to psychoanalytic constructs. The id, ego, and superego reside in this category. They are not physical compo-nents: they are psychical constructs which together allow us to function correctly. The id is an inaccessible element, unconscious in nature, devised of pure desire and propelling the need for instant gratification (thus obeying Freud's Pleasure Principle). The ego moderates the desires of both the afore-mentioned id and the superego, both of which place unnecessary demands upon it. Freud states that it is also formed of prior identifications, "the ego is formed to a great extent out of identifications which take the place of aban-doned cathexes by the id."[5] Therefore the superego might be considered the id's opposite, for it seeks to delay the desired gratification by acting as the conscience. Essentially, the superego acts as the parental voice: that which distinguishes between "right" and "wrong." Within this superego is the ego ideal, a component that, according to Lacan,[6] is formed at the first point of logical interaction with the mirror. In stating something as "logical" here, I mean the point at which a child can look in the mirror for the first time and recognize themselves in the reflection. At this point, the ego ideal (*imago*) develops. This ideal is that which can never be achieved, and remains unconscious. It is the nature of identification. This process, by which the child assimilates her or himself in the mirror, is known as the Mirror Stage, which Lacan described as "a particular case of the function of imagos."[7] This stage will become vital to the concepts I utilize throughout this paper.

The concept of identification is incredibly important, whether we perceive an interaction between two individuals or an individual and another body (such as a video game). Let us then begin to relate it to the game situation. When indulging in this situation, we are faced with a construct which situates us before a screen. On this screen exists the character the player must adhere to throughout gameplay, and a representation by which he or she can move around the world, sometimes called an "avatar." This avatar[8] might take the form of another being, a hand, or even just a pointer. The location of the player to this is interesting, however, because he or she must at once be playing *as* that character but also be identifying with it, as will become apparent later. Players cannot be the avatar but they do assume a level of becoming at the point of interaction. If individuals find themselves sympathetic to this avatar, this protagonist of the game, surely this is enough to allow them to continue playing. As Fine points out, "If a player doesn't care about his character then the game is meaningless."[9]

However, there are times when we do not feel so moved by the actions of the main character in a game. Sometimes we might find that the actions of the character do not abide by our own mindset. In many games the main protagonist is fairly neutral, and so in introducing this "empty" character, we may project our own perspective upon them. One might also include in this the "learning" character, whereby in being new to a world, certain characters reflect our own lack of experience back at us. In short, they become a representative fusion of the self and the gameworld. In a situation where we are involved with a character that is quite different from ourselves, we do not necessarily stop playing the game, however. We form attachments to the game on two levels: that of the player-character or avatar and that of the game itself, both of which are elements that might motivate us to continue play.

How does this relate to identification? The very dynamic of the screen forces us into a "mirrored" situation. The reflective nature of this allows us the chance to form attachments with the character in a way that we might not necessarily otherwise be able to form, and this is where the concept of the intermediate begins to become apparent.

AN INTERMEDIATE CONSTRUCT

The levels at which attachments are formed constitute an important element of play. While attachment is not the entire reason we would continue to play, it is a vital one. I would at this stage like to introduce the construct of the intermediate ego, a psychical intervention based upon psychoanalytic theory which, I propose, allows us to structure our identifications and maintain interest in the product even when we are not specifically at play. This concept will assist our understanding of what might occur when we interact with the game environment. Think of it as a structure that formulates within the triangulation of the user, screen, and game, the purpose of which is to ensure effective communication between these three components and maintain play until the completion of the game. The intermediate ego should be considered an element of the psychoanalytic ego, but one that extends the purpose of the aforementioned ego in a way that allows one effective communication with and understanding of the virtual.

On the surface, the intermediate ego works in much of the same way as the aforementioned psychoanalytic ego. The intermediate is situated in the position of the Imaginary, being linguistic in nature, and the position of mediator between conflicting elements. However, the constructs this ego mediates between are not, as in the case of its namesake, based in the internal world of the individual, for the intermediate takes in and formulates a two-way link between the game and the player. This psychical space thus incorporates the player's internal world, external reality, and the game's "truth." These elements in themselves might be described as so conflictual that we would be unable to even begin play. Therefore, it is up to this construct to assess and mediate the possibilities, thus allowing elements of each of these "truths" to be taken into account without damaging our desire to play. Thus each individual's experience of certain games would change over time depending on the weighting of these elements at the time of play. Tastes towards certain games would also be affected and should predictably not only change over time but differ immensely from person to person. Though we are forced into a situation of identification due to the mirror-like nature of the user-screen dynamic, one assumes that differences in taste between individuals are based upon this dynamic of internal and external spaces. While the game space may "speak" in the same way, the way it is interpreted and understood will differ.

I have briefly shown elements of the game space when the individual is at play, but what is it that causes an individual to play in the first place? Games have a level of desire to them. We desire to play and achieve a level of mastery over a game from the point of initial interaction, which might be an advertisement or seeing it in a store. I suggest that the intermediate ego comes about and is maintained through the psychoanalytic "lack," something we all strive to overcome. Throughout our lives, we are constantly striving towards things that are not achievable, something that is connected with the concept of desire.[10] The lack refers to the lost object, the mother's breast. In losing this object, we spend our lives striving to regain satisfaction, and this is a reason why we might indulge in certain activities, video games included. Bogost states that, "For Lacan, the subject's relationship to the *object* is an impossible one. To seek out the object of desire is not to achieve it, but only to readdress its function and position as a lack."[11] Later Bogost brings the idea of the lack into the realm of simulation, where he specifies these "gaps" as a premise for the creation of meaning. This, of course, does not answer the question of why some individuals attempt to alleviate this lack through playing games while others perform other hobbies and interests. The concept of the intermediate ego goes so far as to describe what is occurring through the duration of play, and while it is possible that it might be effectively applied in other situations, it is not my desire here to describe how it might do so, neither is it to explain why some individuals are more inclined towards certain types of play than others. I do, however, wish to use the construct to show how far it goes in explaining why games are structured in quite the way they are, as we will explore later.

Playing a video game can only be a brief distraction from this loss, but a game exhibits a complex way of offering us two distinct types of gratification: instant and delayed. Instant gratification might be gained from defeating an enemy, winning an important item, or gathering a certain number of points in one turn. Delayed gratification refers to far-off goals, such as completing the game in its entirety or completing a quest chain. In offering these rewards, we are constantly in a state in which we might overcome this loss, but unfortunately finishing the game diffuses this act of overcoming and encourages us to indulge in similar activities in order to rid ourselves of this frustration once more. This may be why recent games often employ generous learning curves for the player: to fight a "boss" that proves too difficult for the player only serves to sever the intermediary link, which only increases frustration rather than alleviating it. The act of doing to overcome becomes one of excess, and the game takes on its own paradox. One might complete the game but in doing so one loses that which was playing a part in overcoming the loss. Completion of a game is in itself a loss, but the strength of the reward that emerges from completion means that we will not immediately need to return to the act of play. I will show later that the game employs the intermediate in order to maintain this interest and prevent us from re-entering a state of lack during the game's play, even when we are not actively at play.

In suffering from the lack, or the loss of an object that can never be regained, we attempt to alleviate the anguish this causes in a number of ways. Interacting with or playing a game temporarily does this by situating us in this space that gives the effect of "between worlds." This space is imagined, but it provides us with the temporary belief that we have countered the lack, which is the important element. Considering the longevity of games, that their playtime can range from anything between a few to hundreds of hours (perhaps even more!), the intermediate formed relating to that game must exist for a potentially extremely long period of time. The illusory space that this occurs within is the location of the identities formed around the intermediate ego, and is based upon the Winnicottian notion of play spaces,[12] as explained later.

I mentioned earlier that games are structured in certain ways, and this is for the purpose of hiding and revealing to us our lack. Games not only alleviate the lack for a period of time, but they also show it to us. This shows the complex nature of the mirroring dynamic surrounding the game, for it not only externalizes the human condition but allows it to emerge into the game space itself. The method of achieving this is through replication, and is a reason why the intermediate construct might be able to survive for such lengths of time. The lack[13] can be shown as replicated throughout the game itself, for it is a position that some games characters appear to occupy, as though to mirror or remedy our own situation. Taylor states something similar, "Game worlds do not lie outside of our ongoing cultural battles, anxieties, or innovations but very often mirror them quite well."[14] The interesting aspect about this is that such characters are not necessarily player-characters. This is a necessary level of proximity for us, and also gains the appearance of us "looking out" at what is occurring rather than perceiving what is happening inside ourselves. This means that the game does not have to damage our internal comfort zone in order to show us what is occurring. If anything, the concept usually takes the form of more borderline characters or ideas. For example, consider the "Heartless" and "Nobodies" in Square-Enix's *Kingdom Hearts* series of games.[15] These creatures take on the role of enemy within the game, representationally causing us to attempt to defeat the lack. Perhaps this defeat is a reason for the existence of so many games whose goals entail one to defeat or destroy other characters or players. Defeat of the other prevents painful revelations within ourselves. We perceive a visual representation of something it seems we ourselves do, and yet we are separated enough from the situation purely by the fact that we play as this figment of impossibility, and this protects us from our actual motive of play. In the Disney-themed world of *Kingdom Hearts*, these collapsed hearts and souls left by their hearts seem somehow reminiscent of human morality. This is relevant because in separating those beings who have some "darkness" in their hearts from those without, we are presented with an impossibility. The pure being that is our avatar throughout the game, Sora, is himself this impossibility (the illusion of ideal, unobtainable, beyond human), and certainly not representative of our selves. Thus, we must see Sora as something akin to the ego ideal. He does not represent our own ideal, but in the situation at which we are interacting with the game, the intermediate ego provides us with a temporary identity which exists as a combination of external, game, and internal "realities." Once formed, this provides us the illusion of ideal. Therefore, one might look at the "Heartless" and "Nobodies" as covering the lack by moving in on the avatar. Yet, as controllers of Sora himself, we are placed in a situation of opposition.

So is it not simply coincidence that this "lacking" character has emerged in video games? Not necessarily, for there are numerous examples of this type of exchange within popular culture. Whether it is the Hollows in popular manga *Bleach*, or those left by the destruction of the hourglass in *Prince of Persia: The Sands of Time*, these monstrous figures emerge time and again, portraying to us an aspect of our nature that can only be revealed in a form quite unlike ourselves, or as a split from the main character.[16] The intermediate ego allows play, yet it also provides the role of adding to the distance from it, thus allowing us to continue to enjoy the play element. Even if the lack is identified in these characters, the intermediate ego rationalizes this occurrence. This, in turn, prevents us from becoming too close to our own lack.

PLAYING IN SPACE

The avatar locates and places us in the location we need to occupy in order to progress through the game and achieve limited satisfaction. Therefore, I explore our interaction with the avatar further, in that a less destructive form of the nature of our defenses is always found in the protagonist player-character. In some sense, this avatar plays dangerously close to our nature, and yet, because we reside at or behind the location of this avatar, it takes on something of a different form. The form it takes depends on our positioning within the game dynamic as players. Sometimes this space is occupied behind the character, sometimes within. Each of these dynamics will be explored further later, with special reference to how this effect is maintained within the intermediate and its importance in these varied situations. The player-character is on a quest to find something, but it continually occupies the ideal, and in doing so the game allows the avatar in most cases to succeed. This is the reward element of playing a game. Even if we cannot gain control over our lack, the journey of this ideal and its eventual end produces satisfaction within us. Yet these phantom stages can only inform us of so much. Our evaluation of the character when we perceive the avatar on the screen for the first time is not necessarily of desire, though it should be considered as something similar. As aforementioned, the player-character is there to place us. Various places within the game itself should be considered as this location of desire. It should not necessarily be limited to the scope of the avatar itself.

Atkins states of the character of the Prince in *Prince of Persia: The Sands of Time* that, "he is not so much representative of the player who plays, with all the flaws and failings of the all-too human player who will fail and will make mistakes, but of some posited ideal Prince."[17] Having not achieved full understanding of the game we are in, we strive to be more in the mindset of that avatar so that we might better grasp the game. Therefore, our identification with the avatar is not necessarily a natural phenomenon associated with our position with relation to the screen but one of forced assurance. In order to play the game, we must identify, however far from our own level of morality our player-character happens to be. Our perceived (forced) virtual body is able to achieve what we cannot, thus we occupy the position laid out for us by the game, which is in turn enforced by the intermediate ego. In this sense, our virtual body is created to appear limitless, and yet is extremely bound: "Which is all, ultimately, an avatar *is*: a perceived world and a sense of control; a moving map, like a projection on the inside of a sphere, and outside it the intention, attention, urgency and passion we bring to our virtual pursuits."[18] It can be switched and molded, and it can end and be reborn, yet it can never truly be our ideal, because in the end it is void of meaning, except to the player, who fills it with this. Of course, this differs with the more direct approach of the player-character who occupies a position in which they may speak freely. However, I would state that a similar process occurs. The speaking character merely occupies a different position within the gaming space. We do not fill this avatar with meaning. Instead, we interpret their actions in ways that seem conclusive for us.

This is much like Caillois'[19] concept of mimicry and taking on identities, whereby the position of knowledge is assumed only at the point of realizing what is behind a mask, at which point one is initiated into being able to use the mask. The intermediate assures that, for the duration of play, such a bridge is in place. In assisting this transferential relationship between the player and game, we are able to externalize the ideal. This is something that sets video games apart from other media, and is based upon activity/passivity and the position of the individual to the screen.

I now return to the space in which the intermediate ego is formed, as it shows how the player as mimic can be initiated into the gaming world's rules. Winnicott[20] theorized the idea of a separate play space which is based on the relationship of trust between the mother-figure and the child, an area in which the child can separate itself from what has been rejected as a part of the "self" (the "me/not-me" dynamic), "This 'space' or gap in-between Self and Other (Other may be a person, people, landscape, object etc.), and or Self and components of Self, has a mediating, relational function where our inner and outer realities can co-exist and from infancy be explored and developed."[21] This concept makes up a large part of Winnicott's extensive play theory, and translates as the psychic space in which the child is able to play. However, if the relationship between the mother and the child is fraught with trust issues, the play space will be ignored. No play will occur. This appears to achieve something similar to the intermediate and therefore to the identification of the player as mimic of the 'between worlds' effect in gaming. Of course, the idea that the player is between anything is potentially a dangerous statement to make, but perhaps there is a play space (this would have to differ from the ludic space projected by the game itself in this situation) in a similar capacity to what Winnicott described. What we must take into account with this construct is that it would require that the intermediate form at this time, but the idea of external relations affecting the duration and ability to play is important here. I would theorize that perhaps this is initially based upon the relationship between the mother and child, as proposed by Winnicott, but later transforms into something far less egocentric and involves external factors such as social involvement. Much like the idea that we replicate aspects of the Mirror Stage throughout our lives, it is also possible that this psychical play space which represents the intermediate forms and reforms throughout our lives based upon our relation to different environments (both internal, external and virtual). The "return to" nature of the intermediate does not simply refer to this return to the past but also a return to the avatar, to the game situation.

In proposing the concept of the intermediate ego, I am taking into account this space (or something evolved from this) which allows us to play. I have already mentioned that the Mirror Stage might be thought of as replicating throughout our lives. In gaming something akin to this stage appears to take place, too. The relationship between the user and the screen could be defined as a return to these stages. Let us not forget that our reason for indulging in things is to overcome the lack we perceive within ourselves, and thus this kind of reasoning has its roots in the idea of the lost object. In "The Lacanian Trio," I return to an important aspect of Lacanian theory as a way to further expose this relationship.

THE LACANIAN TRIO

In Lacanian theory,[22] there are three distinct components that are essential to our understanding of the dynamic existing when the user is at play. These are the Real, the Symbolic, and the Imaginary. The Real exists beyond language; it is something that cannot be theorized with words. The Imaginary is the location of the ideal and is image-based, existing within language as it does. The Symbolic represents the rules surrounding the child and its overall acceptance of those.

In his book *Youth Fantasies: The Perverse Landscape of the Media*, Jagodzinski[23] explains that this Real is somewhere beyond the head of the player interacting with the game. If we perceive his idea that the mirror is doubled, this works in line with the notion of the Lacanian gaze: that is, the object is looking back at you. Here, the screen plays a frame, a window, and a double mirror. Much like in literary criticism, the frame encloses the action, separates it from our being. Equally, the mirrored screen reflects the look back on us through the windowed screen, while also reflecting our own look back (it is, after all, a double mirror). In turn, this draws us closer to the perceived action. However, the look that gazes at us must be false. The look we perceive is interpreted and therefore must still be our own ideal look. We receive the look in return, but that which gives the look is not our own reflection, it must only be perceived as something similar. In much the same way as we are split from our ideal, we are split from this avatar, which does not occupy the same space as the ideal but might be considered to perform a similar role contextually, in the sense of the screen-user interface spanning a gap between the external and internal worlds, which duplicates the look. In the mirror, the look can only be reflected back. In a game we gain the appearance of a duplication of this. Avatars mimic the idea of an ideal, but do so whilst being centered purely in a discourse of play. That is, both reflect a fiction, it is just that each fiction represents a different aspect of identity: one routed in a personal narrative, and the other in a social one.

Based on this analysis, the intermediate ego in this context is the result of the destruction of the linguistic body to a predominantly representational or virtual level. However, I already mentioned that the intermediate is situated linguistically, and thus this clash between linguistic and beyond, between the avatar and the player is replicated through this space. One's body in reality becomes rejected: it is present but exists in flux. This differs somewhat from Jagodzinski's interpretation of "a collapse of the Real into the Symbolic Order,"[24] for the ego in flux does not seem able to rest within the mirror as he suggests, simply between it. Stating that playing a game places us between worlds is, therefore, entirely possible, and the overriding factor that allows us to remain in this position is that of the controller.

The idea of something occurring structurally when we interact with the game's environment is key here. There certainly does seem to be something occurring beyond the level of the intermediate, a kind of passing through of the Real, especially in the case of the controller. This is because the controller not only performs the function of distancing one from the action (as might be seen in such titles as *Black & White*,[25] in which the hand in reality performs actions on the keyboard, which in turn is translated to the hand on the screen) but also in the act of becoming. I have already mentioned the strength of the identification between the user and the game. The controller is an aid in ensuring that this level of projected identification occurs successfully. After all, there is a reason why some games favor distanced spaces above others, and the controller must assist in reflecting the proximity through the intermediate. The gun controller is not a gun in reality, but another controller. However, it mimics the gun on-screen, which is a gun in the *game's* reality. Obviously, this is conflicting. We are in a space of empty action, while the game's reality is in a space of controlled action, so evidently there must be some way to translate us from this space, which is the task of the controller. The gun is nothing more than an extension of the limbs which enables this to occur. Notice also that by mimicking the device the avatar uses on the screen, the gun controller actually brings us closer to the action, rather than separating us from it. I will now expose different types of action in the game setting and different types of game to show how interaction with the intermediate can be produced and strengthened in order to propel us towards continuation of play.

TYPES OF INTERMEDIATE ACTION

I have established some important facets of what occurs during the intermediate ego's presence, and even the location at which it emerges. Why is it so important, and how does this differ from other theories? The intermediate ego emerges from our desire to overcome the lack and allows us the exchange between two realities, whereby at the point of interaction with a gamespace a temporary identity is formulated. Through this "identity," the intermediate ego mediates between the known and the unknown, initiating understanding of objects and acts within the game that represent those from reality. It aligns our purposes with that of the gameworld and grants us the power to play in another world without propelling us into a chaotic state. When at play, we remain situated before the double mirror, the replicated effects of that occurring endlessly throughout the game itself. What propels us to continue to play? As discussed, reward-based systems are in place in games in order to ensure one continues to play until the end, which is connected with the psychoanalytic lack.

Cutscenes, scenes without user-controlled action that are generally utilized to progress the storyline of the game, vary widely but all seem to have the same structural imperative. With the progression of graphics considered "ultra-realistic," much of what we now see in cutscenes is done in-game. That is, the same process that is used to create the play element of the game is also used for the sections in-between, rather than the process of using full motion videos (FMVs), which was previously popular. The reason for this change seems to be a combination of the ability to actually achieve graphics that used to be shown off in the FMVs, and a device to prevent the user from being removed from the space they occupy at the time of playing the game. Reaching a cutscene in a game is a milestone: it states that one has got that far and rewards one for doing so by progressing the plot. However, damaging the play space is not necessarily beneficial to this. Even though the control is removed from the player's hands, they still expect to remain involved in the same action. FMVs render the player powerless, into the realm of cinematics: it is as though they are watching a film. This puts them in an entirely different space to that occupied when they are playing the game. With games which use in-game footage for their cutscenes, the control may be taken and given back to the player at any point in time. The player, thus, exists within a dormant element of the game space, waiting for the control to be returned to her or him (which could be at any moment, because they are not simply watching a video). This drawing back of the action releases the player from the controls but also allows them to perceive more of the game world and think about the situation. It also allows them to see their character in action, whereas before they may have only seen it in first-person, which plays on the doubled mirror, thus increasing identification with the gamespace.

However, there are times when individuals begin playing a game and stop. One might become frustrated over a particularly difficult enemy, or one might become bored with the repetitive nature of the game. Quite the opposite might occur: the enemies might be too easy to kill, or the game might be so random that one spends most of their time getting lost. Perhaps a new game might come out, thus propelling one to abandon their enjoyment in the previous title in favor of a new experience. Because the intermediate ego is fuelled by enjoyment, games designers must be careful to create a game in such a way that players of all abilities can complete the game, but not too easily. Therefore a precedent must be set: the game must challenge, but not too much. Difficulty levels match up to these abilities and ensure that the player is challenged. Yet, if one's desire is to overcome the lack, why not make the game easy? If an individual finds a game too easy, they will not generally play: it presents them with little to no challenge. While the intermediate ego is fuelled and driven by this quest for enjoyment, this must be achieved through challenging means. After all, what is the point in receiving a reward if you did not do anything worthwhile to receive it? As I have already stated, there is a tendency for players to give up on a game if it is too simple. Players are aware that they may have to play a part of a game over and over, that they will be removed from the continuity of the intermediate/ play space because of this. However, playing an aspect of a game for too long can have dire consequences. Less and less time is spent on the game, or the player gives up altogether. This is a destruction of the intermediate identity (please note that this is not the destruction of the intermediate ego, only the so-called "intermediate" as the space created by the intermediate ego. The intermediate ego is always present in some form). At the point that the player loses interest, the almost transferential bond (of which the intermediate) begins to fail. During this time, the player might attempt to rekindle this connection, perhaps by reading guides or discussing the game with others. At this point, the intermediate identity might collapse in on itself altogether or merely dwindle away until the player has finally lost interest. It is possible during this time for another intermediate to form, either alongside it in the same space or after it. This one differs in the sense that the internal and external factors brought to it differ. For example, one might begin playing a game in which something occurs that you do not agree with. The game's handling of that incident might not match up to your expectations based on the external codes of conduct you are bringing to the game. From this, you might feel that the game's handling of that situation is justified given the circumstance and continue to play, or you might be so appalled that the codes of conduct that the game is projecting are so different from your own that you can no longer play the game. The intermediate collapses. At some point before this, you might have decided to play another game. Around this forms a separate intermediate, perhaps one that does not require the same understanding as the previous one did. Perhaps you complete this game, and in doing so decide to play other games in the series. The intermediate in this case would remain in existence for a much more prolonged period of time, and whilst it would not be the same for the next game you played (even if that game was a part of the same series), you would nonetheless retain some elements of that game psychically. These would become useful to the next instalment of that game: one might understand the combat system, recognize certain places or enemy weakness, or even receive a free feature in that game for completing a previous instalment.

Removing oneself from the game space is not specifically a problem to the player, nor indeed to their desire to continue playing it. Once one has removed oneself from the space, there are a number of ways that the intermediate identity formed from this external ego can sustain itself. Staving off pleasure or delaying gratification is one method by which this intermediate remains intact, but this alone is not sustainable without the individual who is not at play thinking about the game, through association such as talking about it or visiting forums. Perhaps the player might purchase merchandise from the game.

Merchandising is the collecting of items related to the game. These items might be apparel, books, special versions of the game, toys, or even random objects from the game. This should be considered as something akin to bringing an aspect of the game into the Lacanian Real. Distanced from the game space, indeed from our own intermediate, merchandise serves as a reminder of one's ability to access that space and prevents one from losing the egoic associations connected with it. If conditions that maintain interest in the product between play sessions are not maintained, the linguistic, imagined element of space created by the intermediate will begin to dwindle. Essentially, the imagined space could be considered forgotten: to replay the game would be to relearn this space, to learn again the language and nature of that game. Some spaces last for longer, such as continuous games that have no proper ending (rhythm action titles, such as *Dance Dance Revolution, Para Para Paradise,* or *Osu! Tatakae! Ouendan* in which one attempts to move one's arms, hands or feet in time with the music to a sequence of increasingly complicated sequences), while more story- or character-based games are much more tightly constrained spatially. [26] To relearn the space is an important component of the collapse of the intermediate, for if one does not have to relearn the space then the intermediate can be recreated: that is, even if the game is forgotten it may still be played again with ease.

This is not the case with Massively Multiplayer Online titles (henceforth shortened to "MMO"), whereby the story continues almost indefinitely. MMOs include games such as *World of Warcraft* and *Eve Online*, which encourage the player to expend a substantial period of time in the game world on quests, developing social links, levelling up, and collecting items. [27] Skills, once learned, must be honed and recalled for future use. There are certain expectations of what one is able to do at specific levels, and to forget these after a period of non-play would be potentially devastating, not to mention costly due to the potential loss of social links connected with it. Individuals invest a great deal of time in these games. Perhaps MMOs are the most pure of games in their ability to continue where any other game would have ended. By constantly adding new content, players maintain interest (an interest already invested due to the quantity of hours already spent) and elements of continuity, the likes of which other games cannot necessarily provide. Yet, players do not necessarily play every day. Some might become restless and leave to play another game, only to return months later. In these situations, I theorize, there is something quite different occurring with the space formed by the intermediate ego. Because of the sheer quantity of hours already invested in such a title, the space formed is more difficult to breach. At the point of beginning the game, the player was aware of the implications of playing a game without end, and they chose to invest that time into that product.

The secondary involvement that maintains the intermediate in this situation is money. In MMOs, it is usual for there to be subscription fees for which players are provided additional support and content, as well as the ability to play the game. In turn, this paying for content serves as a reminder of that game and maintains play. If one is paying for something, one will generally want to receive the best from that item. In the same way, MMO players invest more time and seek out pleasure from that, thus strengthening the intermediate space created. This, in turn, makes it more difficult for the player to stop playing: they have a network of friends, a high-level character and lots of quests left to achieve. Giving that up is far more difficult than deciding not to play a game after four hours of use. An MMO player has invested in their loss, and has become a part of a network who are doing the same. After all, why attempt to find a game you like when the last one is complete when you are enjoying a game with potentially limitless depth?

This brings us naturally to the problem of, when there are games which can maintain an intermediate for such a long period of time, why players would choose to play different games, thus creating a number of "identities," especially considering that in most of these long-term games one can create many "alts,"[28] thus redefining the story depending on how or who they wish to play at the time. As I have already established, however, playing an MMO is both costly and time-consuming. For a solid intermediate to form, one would need to invest heavily into the game, something which many players are not prepared to do. Furthermore, the ending of a game might be considered one of its fundamental aspects. To lose this aspect might be considered unfathomable: to complete a game and then continue the cycle through the beginning of another indulges a new challenge. While the intermediates formed in this case are far more basic than those formed in the case of MMOs, indeed they are more likely to be disregarded, they are also less rigid in their boundaries. Therefore the exchange between the intermediate ego and its intermediate space is more flexible. With MMOs, after a period of play, there is little room for exchange between the two anymore: the game has been understood in a specific way, using certain social and cultural perspectives brought to it mainly at the instance of first play. A game that has not been played for as much time does not suffer from the same problem, as these boundaries are more relaxed, which means that one's perspective might change more easily in terms of a game that promises an ending than in one that does not. I will now cite examples of types of game and show how the subtle differences in the intermediates formed depending on the type of game one is playing. Let us remember, however, that there exists crossover in each of these game "types" and that the following examples are generalizations of what might be perceived to be occurring within the play space the intermediate forms.

God games—games which allow the player to take on godlike powers. Examples of this may be building a garden (*Viva Piñata*), raising a civilization (*Black & White*), or controlling the lives of simulated individuals (*The Sims*).[29] The "avatar" in the game is usually a pointer, an object, or a hand, and this is what reflects the desires of the individual and translates the actions into the game. The game is usually controlled with a keyboard and mouse, although more recently there have been ports of such games to consoles, whereby a controller might be used. Distance is maintained by the translation from hand to screen, the fact that the avatar is not an animate object (identification is limited because the motives are left to us) and because of the top-down view (the game is generally viewed from bizarre angles, such as above the action, rather than us being in the center of it). In this case, the intermediate forms in perhaps a more difficult way to that which has already been described, for there is little room for identification with the avatar. We need to perceive the fact that the game wants us to be situated in the position of the avatar, which is invisible. This shifts our proximity to the screen, and perhaps even makes this identification all the stronger. We occupy something of a perverse position in this scenario: it is as though the game places us in the position of the lost object. Because of the duplicity of the screen, however, this is not feasible, and we are left in a space as close to the "between worlds" effect as is possible given current constraints. This intermediate is maintained through the empty or invisible avatar. Completion of the game renounces one's godlike status, thus collapsing the formed intermediate.

First-person shooters (FPS)—FPS games (e.g., *Bioshock* and *Halo*) initiate the gamer once again into keyboard and mouse territory (though, again, a multitude of such titles have been ported to console systems).[30] Occasionally gun controllers are also used, such as in *House of the Dead 2*.[31] However, FPS titles are all about proximity. One must be as close to the action as possible at all times. This is achieved through the use of an avatar that one "drives" as such. We are situated within another almost perverse space, within the avatar, for all that we see is their arms. When we press the "jump," "run," or "shoot" keys, we see the avatar as performing those actions within that space. So, even though the keyboard and mouse should serve only to distance us from the game, we are actually drawn in, possibly due to the directness of the action. There is nothing further to select once you have pressed a key (with the obvious exceptions of changing weapons or saving), and being situated within the avatar space certainly assists in this. A more comical way of drawing us in further is in the game *Typing of the Dead*[32] where the avatar does not hold a gun but a keyboard in a direct mimicry of our own action (the aim is to type the words listed on the screen in order to perform attacks). The keyboard thus becomes the weapon. In this case, the intermediate forms in a similar way to as it would in God games, but with subtle differences. We have switched from the position of "god" to "ordinary." Whatever our powers within the game, our control of the environment is ultimately more limited. The space is far smaller, which in turn limits the replication of this space afforded to the intermediate. Imagine the intermediate as being representative of the perceived space and proximity.

Third-person action games—here I refer to such games as the *Resident Evil* series or *World of Warcraft*, in which we clearly see the avatar in front of us.[33] Like with FPS titles, the avatar will often be humanoid, but like God games some distance is maintained between the avatar, screen and player/ controller (I group the player and controller together here, as one might be seen an extension of the other). The controller in question here could be anything already specified. In playing a third-person game, one achieves the effect of slight distance. The avatar is our puppet, and through them we can spend time exploring. Because of this emphasis on devices other than instant action, the avatar is representative of stepping back from the game and working on clues, or looking at the whole area before proceeding. The intermediate in this case reflects this larger quantity of space, thus affording us the added freedom of movement afforded the avatar (and thus the player) in the game. This is actually represented by a larger space than in God games, as movement in such games is constrained. Third-person games rarely exhibit small spaces, and the movement of the avatar is representative of this. In this large space, we are distanced from the avatar. This allows us room to perceive the avatar in a "me/not-me" context, whereby at once the avatar is the player as the player is controlling them, but at the same time the avatar is not and therefore one must almost guess the motives of the character. Further, one might equally be able to witness scenes quite outside of the character, thus offering the player additional information that the protagonist might not be aware of. This type of game can be more problematic in terms of identification as the avatar is often not silent and is freely able to express their opinion in the game. In first-person games, this control is left with us: the avatar can be what we want them to be (within the context of the game). Therefore, the intermediate may be destroyed during the game by loss of identification with the avatar.

In all three examples, we have seen how the intermediate might be created, maintained, and destroyed. Whether it is for the perception of in-game events or simply a feeling of relative proximity to the character, the game manipulates the psychical space formed between it and the user, thus manipulating the effect on the user. This creates a bound yet individual experience of the game because of the influence of the external and personal histories surrounding the user. Thus, I suggest that the manipulation of the game on the user serves as a basic illusion offered to the player as "control." In turn, this illusion can be compounded into a rounded game experience for the user which allows them a level of temporary freedom from their personal situation, whilst exposing them to veiled reflections of occurrences within said reality.

CONCLUSION

The position of the avatar with relation to the player/controller has a significant effect on how we perceive the game we are playing, and indeed how the intermediate forms and functions. While the intermediate maintains the level of play experience, different types of games present it with different problems to overcome in order to allow the player to enjoy the game throughout. Perceived proximity to the action dictates how we will play the game, because it carries a linguistic code of its own.

The intermediate in these situations would incorporate different elements from external reality in order to correctly align with the game itself. As aforementioned with regards to MMO games, this can be a flexible process or something far more rigid, depending on the longevity of the game amongst other factors. However, this does not detract from the sense that we propel and feed into the intermediate different elements from the external depending on the game's focus. This eases the chance for broken identificatory bonds, because if something in the game is blatantly out of balance with the morality that the player has brought to the game, they are more likely to stop playing. However, if that element is a minor one, and not one which we have anticipated at the point of the intermediate space forming, the likelihood of acceptance of such an element from the game is more likely. Therefore, we are more likely to accept what is not considered within us problematic, but this is elevated to the level that we will accept it if it is not an anticipated element within the game. We are always free at the point before initial play not to play the game if we feel that such elements will not match up to our own morality. Yet, it is more difficult to stop playing at the point of this realization as connections have already formed.

The game is already having an identificatory connection with the user before the point of sale, especially with reference to games series. This connection can begin even before the point of sale, perhaps at the point of initial advertisement, and is connected (as explained earlier) with desire and the lack. However, there is need to explain the process that occurs beyond this to propel the user to the point of the game's completion. That is, the user is already playing, and it is understood how they got to that situation.

The intermediate ego, as an aspect of the psychoanalytic ego, acts as a mediator between the individual's reality (including external interventions) and the gamespace (including cutscenes and other elements outside of the user's control). The difference between this egoic space is that it exists outside of the psychic space occupied by the id, ego, and superego, and only really comes into play at the point of intervention with the gamespace. Despite the predominant interest in video games here, this is not to say that it does not occur during the process of playing any other variety of game, although specific aspects would differ, such as the lack of screen as mirror which is central to the idea of play with regards to video games.

The intermediate ego is thus a disposable phenomenon, active only at the time of (potential) play. It is an ego combining physicality with virtuality, and I have proposed that this is what prevents us from becoming the avatar. We are merely in the process of doing so, but this is a process that never reaches completion, for once the game is over we simply shed this intermediate identity and take on another. The intermediate ego is therefore the product of the collision of the game world and reality, granting it the appearance of "between worlds." The intermediate ego allows for the user to identify with a mind and/or situation that they would not otherwise necessarily be able to identify with under other circumstances. It offers a valid reason for doing something. We can perform an action without the guilt or legislation associated with it because we are not, and never were, that particular character. The character and scenario are fictional, and the game world reminds us of this constantly with status bars, save screens, control tutorials, and breaking of the fourth wall. When we switch off the console, the game will not continue to run in our absence: it requires the user to run. However, it is also directive of the user: the user assumes control but actually has very little, as previously established. Sometimes it tells us things where we are assumed to be able to recall similar situations in our own society, in which case it becomes a commentary on reality. What looks back at us is our reflection, familiar and yet slightly warped. This relationship can be explained using "Freud's Death Drive," in which one is constantly propelled towards non-being, as explained below.

When one starts playing a game, the external world "is" and the game is "not." However, immersion in the gaming situation reverses this process. While in a state of immersion, the "not" is the external world and the "is" is the game. After the establishment of the intermediate, these roles bounce between the two, something dependent on the level at which immersion is retained.

Of course, in actuality, both forms exist at once. It is merely our perception of them that changes: they cannot both be envisaged as "real" at once in subjective reality. Only one constant "is" must exist at one time, even if it is a changing process. This idea might be applied to other disciplines, and one can certainly see how similar processes might occur when we conduct other practices, such as reading a book.

In this paper, I have shown the formation of the intermediate ego, how it propels us through a game, and assists us in overcoming our own lack. I have also shown how it assists in manipulating the space between the user and the game structurally and linguistically in order to allow the player to enjoy a more complete experience, and how the nature of the avatar is also reflected in this. I have abandoned the idea of the "between worlds" effect in favor of a psychical mediator which allows us access to a world that directly takes from and mimics our own, both on internal and external levels.

The concept of the intermediate ego has been developed with the purpose of assisting in other studies, to assist individuals in understanding what might be occurring structurally when they indulge in playing a video game. Effects studies consider what is occurring on the surface when at play. I believe that the reality of the situation is in fact far more complex than what we give the appearance of doing during play, and therefore one can only go so far into describing what is occurring when an individual is at play unless one begins to understand the processes underlying this. While the intermediate ego is a purely psychical construct, it is my hope that it will assist in explaining and understanding our interaction with the gaming space in future.

NOTES

1. Bob Rehak, "Playing at Being: Psychoanalysis and the Avatar." In *The Video Game Theory Reader,* ed. by Mark J.P. Wolf and Bernard Perron (London: Routledge, 2003), 103–128; Jan Jagodzinski, *Youth Fantasies: The Perverse Landscape of the Media* (New York: Palgrave Macmillan, 2004).

2. Slavoj Žižek, *Violence* (London: Profile Books, 2008).

3. Amanda Lenhart et al., "Adults and Video Games" (Pew Internet and American Life Project, 2007), accessed at http://www.pewinternet.org/Reports/2008/Adults-and-Video-Games/1–Data-Memo/01–Overview.aspx?r=1.

4. Roland Barthes, *Mythologies*, trans. by Annette Levers (London: Granada Publishing Ltd., 1973), 129.

5. Sigmund Freud, "The Ego and the Id and Other Works." In *The Standard Edition of the Complete Psychological Works of Sigmund Freud Volume XIX (1923–1925)*, trans. by James Strachey (London: Vintage, 1961), 48–59.

6. Jacques Lacan, "The Mirror Stage as Formative of the I Function as Revealed in Psychoanalytic Experience." In *Écrits*, trans. by Bruce Fink (New York: Norton, 2006), 75–82.

7. Lacan, "The Mirror Stage as Formative of the I Function as Revealed in Psychoanalytic Experience," 78.

8. Here, I refer to the term "avatar" as a structural/visual representation of the player on the screen. Though this term may be linked with the idea of the "player character" at the point at which the user is interacting with the game, this is only the case where this representation is representative of the player (which I would argue is at the point at which the player identifies with the game). The "player character" is linguistically represented in the game and may still be utilized structurally as an avatar, but the traits issued to it by the game will retain a level of separation from the user.

9. Gary Alan Fine, "Games and Frames." In *The Game Design Reader: A Rules of Play Anthology*, ed. by Katie Salen and Eric Zimmerman (Cambridge, Massachusetts: MIT Press, 2006), 578–602.

10. At the point the child becomes or perceives her or himself as separate from the mother, the idea of being whole becomes unachievable. See Lacan, "The Mirror Stage as Formative of the I Function as Revealed in Psychoanalytic Experience."

11. Ian Bogost, *Unit Operations: An Approach to Videogame Criticism* (Cambridge, Massachusetts: MIT Press, 2006), 32.

12. Winnicott's notion of play spaces comprises separate psychical spaces formed from a bond of trust between the child and the mother. The child is able to play in this space, providing the relationship with the mother is maintained. I theorize here that we always retain a space like this, but that its formation and maintenance differs from Winnicott's conceptual space.

13. As previously referenced, our loss of the maternal space, and therefore our initiation into the world proper, shows us as incomplete beings who are always seeking to reconnect with this lost object. This exists as an impossibility, however, and all we may do is serve to overcome it through attempting satisfaction for short periods of time.

14. T.L. Taylor, *Play Between Worlds: Exploring Online Game Culture* (Cambridge, Massachusetts: MIT Press, 2006), 129.

15. In Square-Enix's *Kingdom Hearts* series, the Nobodies and Heartless appear as beings that have lost their "hearts" (the term "soul" is perhaps an accurate description of what this is). Heartless consume hearts, while Nobodies are what is left when a Heartless has consumed a heart. This seems fitting: the Heartless are constantly seeking to overcome the lack through consuming it/standing in its place. The Nobodies have no memory of what it is they are lacking in the first place.

16. Tite Kudo, *Bleach* (San Francisco: Viz Media, 2001)—A manga in which a boy must become a "reaper" in order to save his town and his friends; Jordan Mercher, *Prince of Persia: The Sands of Time* (Montreuil-sous-Bois, France: Ubisoft, 2003)—Breaking the hourglass releases the "sands of time" which turn individuals into monsters.

17. Barry Atkins, "Killing Time: Time Past, Time Present and Time Future in *Prince of Persia: The Sands of Time.*" In *Videogame, Player, Text,* ed. by Barry Atkins and Tanya Krzywinska (Manchester: Manchester University Press, 2007), 237–253.

18. Bob Rehak, "Of Eye Candy and Id: The Terrors and Pleasures of Doom 3." In *Videogame, Player, Text,* ed. by Barry Atkins and Tanya Krzywinska (Manchester: Manchester University Press, 2007): 139–156.

19. Roger Caillois, "Simulation and Vertigo." In *Man, Play and Games,* trans. by Meyer Barash (Urbana: University of Illinois Press, 2001), 81–97.

20. Donald Winnicott, *Playing and Reality* (New York: Routledge, 2006).

21. Amanda Bingley, "In Here and Out There: Sensations Between Self and Landscape," *Social & Cultural Geography 4,* 3(September 2003): 329–345.

22. Lacan, *Écrits.*

23. Jagodzinski, *Youth Fantasies.*

24. Jagodzinski, *Youth Fantasies.*

25. Lionhead Studios, 2001.

26. *Dance Dance Revolution* (Konami, 1998)—A game in which one must step on arrows in time with the music; *Para Para Paradise* (Konami, 2000)—A game based on learning routines in order to follow arrows in time with the music; and *Osu! Tatakae! Ouendan!* (Inis, 2005)—A handheld game in which one must trace dots and patterns in time with music.

27. *World of Warcraft* (Blizzard, 2004)—An MMORPG set in a fantasy world with warring factions; and *Eve Online* (CCP Games, 2003)—A science-fiction MMO in which one must customize and pilot a starship.

28. Here, I am referring to the idea of a player having multiple characters on the same account. A player may have a "main" (their primary character) and a number of "alts" (secondary characters on the same account which are used less frequently).

29. *Viva Piñata* (Rare, 2006); *Black & White* (Lionhead Studios, 2001); and *The Sims* (Maxis, 2000).

30. *Bioshock* (Irrational Games, 2007)—A first-person shooter set in an underwater city. This game incorporates elements of role-playing games such as the ability to level up certain elements; and *Halo* (Bungie, 2001)—One of the first successful FPS games on a console (rather than PC) format.

31. Wow Entertainment, 1998. A zombie game in which one is on "rails" (led around the map to where one must fight over a specific period of time).

32. Wow Entertainment, 1999. This follows the same structure as *House of the Dead 2*, with the player typing words in order to defeat enemies rather than shooting them.

33. *Resident Evil* (Capcom, 1996)—Termed a "survival horror" game in which one must defeat zombies in order to progress; and *World of Warcraft* (Blizzard, 2004).

Chapter Four

Producing Place and Play in Virtual Game Spaces

J. Talmadge Wright

Playing computer and console games, as one of the leading forms of enter-tainment for young people and adults, has generated new social relationships, improved existing friendships, and created new venues of play. [1] According to a study by the *Pew Internet & Life American Project* in 2003, 70 percent of college students reported playing video games once in awhile and 65 percent reported being regular game players. [2] Out of 1162 students attending 10 Chicago area institutions of higher learning, one out of five students interviewed felt that gaming allowed them to make new friends and improve existing friendships. Sixty-five percent claimed that gaming did not deprive them of family or friends. And, indeed, gaming is well integrated into other college activities, including listening to music, watching TV, sports, etc.

The social activity that surrounds digital game playing, especially the playing of massive multi-player online type games, like *Ultima Online, World of Warcraft*, or *EverQuest*, has been well established. [3] Indeed, social activity is widespread in all digital gaming activity, despite the public fear that playing computer games will induce social isolation. Even multi-player first person shooter (FPS) games such as *Counter-Strike, Battlefield 2*, or more recently *Call of Duty: Modern Warfare 2*, thrive on the social interac-tions that playing them create, generating entire industries devoted to fan competitions and fan game modifications. While FPS games constitute a minority of those who play computer games in general, studying how those players make sense of what they do, where they do it, and the pleasures they derive from that activity is essential to advancing the debate over play, new media, and social networks.

Data for this study were collected using participant observation and in-depth interviews with 23 players of the FPS game *Counter-Strike* at a mid-western university campus between February 2001 and September 2002. Noted were diverse patterns of game entry, game participation, new friendships, and modifications of physical space for game playing. Throughout this period, it was observed that game players utilized existing friendship networks and created new social networks as they became involved in game playing. In generating new forms of solidarity, these game players took the form of "affinity groups" identified, in part through their status as students but also through gender and age.[4] The fact that all these players were men, with the exception of one female player who played once with the group and another female player who was interviewed separately, also points to the socially constructed gendered character of the genre and the manner in which gender play was constructed through both the character of the play, the talk around the play (from trash talking, reactions to cheating, and the telling of after game "war stories"), and the organization of the play activity itself. Thirteen players were observed playing in two distinct campus locations with ages ranging from 20 to 34 in one location (the library group) and 21 to 22 in the other location (the computer lab group). In addition, 10 other players were interviewed who had played the game in various other locations, some as far away as Taiwan. The ages of this group varied between 18 and 23. All players interviewed, with the exception of two, were men.

Understanding digital FPS games as simplistic point and shoot "games," encourages us to think in a traditional format of rule bound play ignoring the complex reality of virtual worlds.[5] Such worlds of play include both Jenkins' complex spatial narratives as well as what Juul would call "stylized simulation."[6] An FPS game like *Counter-Strike* is, in fact, a game within a pre-defined virtual world, with tactical rules and simulated physics, where player collaboration and game immersion provide the foundation for generating social meaning between players. This willing collaboration defines a social order that is always in flux as players enter and leave the game, change their orientation towards the social rules of the game, their understanding of each other's behavior, and altering the physical setting of the computer's location.

CREATING A SPACE FOR GAME PLAY: *COUNTER-STRIKE* AS A
SOCIAL AND PLACE MAKING ACTIVITY

While game playing occurred in a virtual world using computer generated avatars, the players sat in a wide variety of physical locations. In my sample, the actual material space of game playing was divided between a university computer lab with rows of computers and the after-hour library offices, which included one small office containing five computers and a larger anterior office space with separate office cubicles. The younger undergraduate players occupied the lab located across campus, while the older male graduate students played in the library offices. One of the graduate students, who worked in the library during regular hours, had access to the offices and installed the game on many of the office machines, using one as a server to host the game. The undergraduates appropriated a row of computers in the computer lab playing the game without sound and occasionally attracted attention from other students when their excitement got the better of them. Even though the lab was supposed to be off limits for game playing, supervision was minimal and playing was tolerated as long as other students were not disrupted in their studies.

After the library offices closed for the day, game players entered through a side door at around 9 pm on Wednesday nights and proceeded to the back offices taking over one of the office computers and usually playing until midnight. This library space divided between a small office with five computers, allowing close contact between players in the same room. And an array of scattered computers on the main office floor made for some interesting social interactions as a result of the distances between players. These latter computers in the main office space were perceived as the least desirable to play on, since they did not facilitate close interpersonal contact. Hence, there was always competition for who would get the computers in the small office. Latecomers would be consigned to computers in the larger room. According to one player, Bruce, "I've played it before where I'm either, sort of, physically isolated from the other players, um, the set up . . . (in the library office) is that, there is only about five computers in one room. And once those fill up, some of us will play in a different area. And, I've always found that whenever I play by myself, I, without fail, don't have as much fun. . . . It's fun to blast at one another and sort of carry on as a group activity."

The "library social group" was well established before the actual playing of *Counter-Strike*, with players knowing one another through their involvement in the university's philosophy graduate program. Other players were recruited as time went along, often from within the same department but at different stages in their graduate careers. The point is that the actual physical space of playing was significantly important in shaping the social responses between game members, as well as their actual perception of what gave them pleasure in playing the game.

For the undergraduate players the computer lab provided a social location to meet new players and make new friends. Since most of the undergraduate students lived in the campus dorms, the development of *Counter-Strike* game

playing on a regular basis became the foundation for cementing new friendships. It also pointed to the lack of good social spaces where social networks
could be formed around the campus. Omar made this quite clear,

> In my experience at [midwestern university], there is a great lack of any social
> activity during the week and the student union is badly managed. Most of us
> started playing these games because we found a similar interest that engages
> several people at a time. We were several groups of friends (different majors
> like business, science, computer science) who were drawn together because we
> saw some (one) play and then joined a game. And we eventually over the next
> two years grew to know most of the people in our group pretty well.

Another undergraduate student, Tony, echoes these complaints about the lack
of student social space, commenting that, "there's nothing to do . . . during
the week," adding that, "the neighborhood's not the greatest." Comparing his
campus with that of his brother who was attending a larger university he
mentioned that at the bigger campus there was a more diverse array of
activities. He claimed that he would have been less likely to get involved in
game playing and would probably only use the computer lab when he had to
do his homework if more social options were presented on his campus.

This stark absence of "third places," or informal, comfortable social
spaces where social interaction is nurtured, reflected a poverty of campus
social life which game playing helped to reduce by generating new social
relationships.[7] Clearly for most of the undergraduates, and indeed for the
graduate students as well, the university had to be more than just a location
for classroom work and library research. As Rebekah Nathan revealed in her
study of university social life, the quality of social spaces and the intensity of
social life on campus is often what makes the undergraduate experience
memorable as well as facilitating intellectual discourse and reducing feelings
of isolation and alienation.[8] Absent places for students to meet and "hang
out," intellectual discourse and the free play of ideas is reduced to limited
classroom discussions and lectures, with most students retreating to private
concerns and self-imposed isolation at the end of their classes.[9]

For many students, college dorms provide a degree of autonomous social
space, often for the first time away from home. Even with an absence of
viable public play space, students will often find ways to generate such
spaces where they live. Rafael, for example, even though he was not playing
with our library or lab group, mentioned his experience in the college dorms
at a neighboring college:

> At one time our whole floor played, we would have 6–10 people in one room.
> It just seems that everyone would jump on one team and we had a clan . . .
> basically we had four or five rooms that we played in. . . . We would log on,
> get into the same room and just basically go at it from there . . . with *Counter-*

*Strik*e one person would take control and it was we go this way or that way. It was a lot easier because you could just tell the person next to you, well let's go this way or that way and you would both go that way. . . . We had in the gaming community I guess you would say, one clan on our floor, and then across campus there would be another clan.

What became apparent in my sample was a small group of game players who, in fact, appropriated what the university defined as instrumental "work spaces" converting them into non-instrumental "social play spaces" if only on a temporary basis. This activity, while tolerated was frowned upon by the university since it was felt to be disruptive to other students and a violation of private office spaces. However, the lack of true student social spaces compelled this group to resist in a manner that ultimately facilitated new friendship networks.

Other undergraduates learned about *Counter-Strike* through friends outside of the university and from previous high school friends who were playing earlier games like *Quake II*. With family and friends often providing the early entry point to game playing it is not surprising that trying out new games could be an opportunity to demonstrate one's digital combat skills as well as being wowed by the graphics of the new games. Lewis had previously played the game *Half-Life* and he had played the mode, Team Fortress which he "didn't like . . . too much." However, after downloading *Counter-Strike* from the Internet he said, "'Oh my god, that gun is so realistic,' that was my first reaction, 'Wow, look at all these weapons, they're so realistic.' Eye candy was the real lure. . . . I was like, 'Okay, let's see how it is,' and then I played multi-player." When he first tried playing against other players he discovered that using the same tactics he often employed in playing single player games, did not work and he had, instead, to learn new strategies if he was going to be successful at the game. On the one hand, Lewis wanted to play the game, but on the other he wanted to go back to a game that was less challenging. Given this context, his friendship networks became essential in bridging this ambivalence by giving him friends who being in close proximity could help him through the learning process.

Some of the undergraduates learned about the game through their on-campus friends or dorm roommates. Tony, for example, discovered that his friend had transferred to [midwestern]. In the past, he would look askance at computer games, favoring consoles. However, one day he saw his friend at the computer lab, playing with other students:

These guys would take up the back row and, you know, I'd see him playing all the time, and I'm like, "Man, this looks fun." . . . So, then I slowly started learning . . . with them, and then . . . everyday, I found myself in the afternoons, instead of watching TV or wasting my time sleeping at my apartment, I'd be there with these guys, playing, you know, like, there's nothing to do around here in the afternoon during a school day.

For the older graduate student players, the issue was less a lack of something to do or adequate social spaces, and more one of expanding and developing friendship networks which had already existed, in part, due to their common experience within an academic department. The organizer of the group, Micky, had received a copy of *Counter-Strike* from a friend. Prior to this he had played *Quake* and then *Half-Life*. Recruiting some of his friends in the graduate program he was enrolled in, Micky managed to set up one of the library office computers as a server which would allow his friends to play a LAN game. With several eager game recruits, others began to join the game within the same friendship network.

Jose, who also worked in the library with Micky heard of the game from friends, said that playing with his friends increased the social solidarity between them as a group:

> When I heard . . . my co-workers were playing, and they were always talking about it like they were hiding something, because they were not supposed to be playing, you know, this computer game at work or, where we used to play, inside the library where I work. Um I was interested by the fact that it was kind of hidden and by the way they talked about it was exciting. And so I asked them and they told me about this game *Counter-Strike* and I started playing and said it's really, it's really fun and I like it a lot and so I kept playing, they invited me to play with them and I stayed.

Bruce who also worked in the library started playing with Micky on the earlier game *Half-Life* and then *Team Fortress* about a year and a half before playing *Counter-Strike*, improving their playing skills before graduating to *Counter-Strike*. Another player, David, mentioned that he began to play in the library as well, because he could not play from home without a cable modem. David, in turn, recruited Bryson, who had learned to play *Half-Life* and then teamed up with the other philosophy graduate students to play *Counter-Strike*. Markus, a long time fan of computer games, starting with *Doom*, was recruited by several of the team members including Micky. Tim came along, as well as having been recruited by Micky. As he said, "basically he just asked me to come along and play. And I had played, I mean I played computer games off and on for years. But, yeah, he said he had this group going on Wednesday nights and pretty much it was a lot of my philosophy buddies." Tim, after playing awhile ended up in a dispute with his wife over game playing and quit the group. The other members of the group knew that he had stopped playing and they didn't push him to join. But, one day he was, "hanging out with Micky and he just said, we're still doing it do you want to come by sometime. So, I wasn't all that busy at night so I'd stopped back." Clint was also invited to play by the group after hearing the members joking about game playing.

In the above cases, game entry happened with friends and relatives, but the social and physical location of game playing were essential in order to set the stage for player interaction. Without these spaces converted from their normal work usage to play usage, it is doubtful if playing *Counter-Strike* would have had the same meaning for the people involved. Play spaces are important, not just virtually, but physically as well. For someone who initially did not have access to the Internet, early internet cafes provided a social space for play which, only later, was expanded to the college computers when Internet connections became a standard tool for classroom work. Peter, for example, began playing digital games with his friend on computers in the early internet cafes because he did not have a computer with a network connection at home. He said,

> We started getting into network computers, my buddy started playing on those, and I started playing with him. . . .We just started checking out all the multi-player online games that we could play and of course first person perspective shooters are the big ones, and uh . . . I started off with . . . *Quake II*, *Quake III*, um, *Half-Life*, that sort of thing, and then *Counter-Strike* was sort of the natural evolution to that, it's one of the best ones that was released, and we all just sort of moved on to that.

For Peter, aside from the Internet connection, the reason he played *Counter-Strike* at the local Internet cafe was because it was where his group of friends at the time would gather. As he got older, went to college and his high school friends dispersed, the allure of FPS online games wore off. He said, "And that was really all it took for me to stop going to Internet cafes and playing online shooters, I mean, I've met some new people in college now and they were talking about, you know, getting a group together and going out to Internet cafes and to do, like, a night of gaming, and I'm going to be definitely with them on that." Peter's interest in the game is less about the actual pleasure of the game itself, and much more about hanging out with his friends in a collective setting.

People's experiences with game playing are quite diverse and thereby cannot be captured through controlled experiments, or even narrowly defined longitudinal studies. Human experience is simply too complex to accommodate simplistic notions of game meaning or game entry. Tina's experience as the lone woman *Counter-Strike* player that our team interviewed is instructive, both as a Caucasian female player, but, also as an American player, learning how to play in Taiwan. In the beginning, she mentioned that she did not like playing video games. She said,

I didn't like video games to begin with. I grew up with my dad and my brother playing them in the next room. It was right next to my bedroom and I would go to bed with, uh, I don't know what they were playing, *Civilization* and a lot of other things and I kept hearing, you know, the Borgs say stuff (*Star Trek* game). . . And I was just like, "uh, I hate this," you know. It seemed so mind numbing.

Her boyfriend who had been playing the game for a year in Taiwan returned and told her about playing *Counter-Strike*, which at the time she, "felt it was kind of stupid. I was just like, 'yeah whatever, you know, it probably wouldn't interest me at all.'" However, when she went to Taiwan to work teaching English, she discovered that all of her male friends were playing *Counter-Strike* in the local Internet cafes. She said, "all my girlfriends just did other stuff. Some of my guy friends invited me to come along to play *Counter-Strike* with them and they decided to teach me . . . so they said they were going to train me really well to be their little protégée." This paternalist attitude on the part of her male friends was accompanied by respect for her eye-hand coordination, which had already been sharpened through extensive piano practice. She commented,

They thought they'd train me so that I could really, um, kick Chan's butt basically when he gets back . . . the following summer when he came back I was really good at it. And I didn't really mention anything about it in that next semester, I was just like, yeah, you know, I liked it, it was kind of fun, but not my thing, you know. . . . I don't know if he felt like his territory had been taken over or something, there was a breach of power, I don't know what it was, but, um, he was kind of funny for the first couple of days about it. He was just like, "Yeah, I let you win." You know, that kind of thing, "I let you beat me."

After recovering from this initial challenge to gender power divisions, both Tina and her boyfriend, Chan, went on to form a gaming partnership using feminist names when they played with other players.

Because Tina's boyfriend was Taiwanese he blended in well with others who played in the internet cafes. However, Tina's status as a Westerner, American, and white female placed her in a confusing position with local gamers who understood Westerners as using computers primarily for checking their emails. She mentioned,

They didn't react to me as well as they reacted to him. Because, first he was Taiwanese and he looked Taiwanese, so, they had . . . no problem with him blending in. Um, but then, for some reason the westerners that came into the internet cafes didn't usually game, they used it mostly just for . . . emails and whatever, cause they were just in town for six months or something . . . teaching English. . . . So, they didn't really want to interact with many of the Taiwanese students . . . most of them (gamers) were then shocked that the person that they were playing with that was beating them was a girl, you know. And because I was . . . using Taiwanese or Mandarin speaking on screen, they didn't identify me.

Tina's understanding of the gender politics of playing *Counter-Strike* was accompanied by how those politics were negotiated within the setting of the Internet cafe. As she pointed out:

> It was huge Internet cafe, so you could be sitting anywhere and they would have no idea, you know, who you were playing. Whereas in the United States, most of the ones I've seen are like, ten game machines and that's about it . . . so you know who you're playing with, and which person they are onscreen just by looking at their face or hearing them actually talk or whatever. But there [in Taiwan] they don't actually know so they would just think I was a guy for a while, and then after a while I'd say some things that let them know I was a girl, you know. Like, if somebody sprayed internet pornography I would cuss them out, you know, or something like that. . . . They were always just kind of like, "Why do you have a problem with this?", you know. And I'd be like, "Cause I'm a girl," and they were like, "you're a girl?" And they would say they didn't believe me, you know, so I would stand up sometimes and be like, I'm the one over here. And everyone would see me and they'd be like, "and you're white too?" . . . And so, for a while, a lot of people didn't quite know how to respond to me, you know, as, like, good CS player who was also female and white so it was kind of strange, and who also spoke the language. But then after a while, quite a few of the regulars kind of knew me pretty well and so we'd play together.

This collusion of culture, language, and social space reveals the complex fashion in which play identities are negotiated not only in the act of play itself, but, also its settings. It is fashionable to talk about how virtual spaces are really not that far removed from real interactions in physical space. Our work seems to confirm this for the participants. In-game behavior often paralleled off-line behavior.

THE SOCIAL PLEASURES OF *COUNTER-STRIKE* PLAY

The social pleasure of playing MMORPG games, like *EverQuest*, and FPS games such as *Counter-Strike*, is re-affirmed in the early work of Stephen Kline and Avery Arlidge. In their survey of 1,178 computer game respondents, 85.7 percent expressed a preference for role playing games followed by shooter games at 75.9 percent. An overwhelming 87.5 percent of participants mentioned that communicating with other players was an important factor for their involvement, followed by exploration of the game environment and player teamwork. [10] While the survey included both lovers of role playing games and shooter games, it is still possible to break out what about shooter games appealed to online players.

Interestingly, *Counter-Strike* players actually practiced less communication about game tactics while playing than did players of *EverQuest*. It is not surprising that exploring game spaces, enjoying plots, themes, graphics, and cooperation would rate very high among this sample given the dominance of role playing games compared to shooter games in the market. However, even so, unpredictable game play (68.9 percent), game play that makes you think (66.9 percent), and feelings of control while playing (66.8 percent) were well established for both game genres. It is also not surprising that players of *Counter-Strike* were overwhelmingly male given the dominance of gender socialization patterns which force both genders into pre-determined stereotypes, "Only 17.6% of female respondents have tried the game, compared to 59.1% of males, and only two female respondents in total claim to be of average skill. This is supported by the findings that 24.8% of males strongly liked "Fighting / Shooting" games offline and 46.3% online, while only 4.2% of women rate fighting / shooting games that well regardless of whether they are played online or offline." [11] Hence, the experiences of Tina mentioned earlier still appear as an exception to the "norm," even with the expanding importance of women who play FPS games.

The undergraduate players in our sample were more likely to engage in online competition with other *Counter-Strike* players, while the older graduate students expressed a preference to play with their own friends through a LAN system. Therefore, the actual physical setting of game play was very important for the older players, a setting where they could talk with one another physically while playing the game. This was less significant for the younger players. Having play spaces in close physical proximity can facilitate social interactions. The development of Internet cafes that encourage social interaction and a relaxed atmosphere, like the back offices of the university library, can indeed provide a context for everyday social interactions, as evidenced by the expanding public gaming sites throughout Asia. The expanding role of mobile media and the increasing use of voice-over programs like Ventrillo or Skype expand the possibilities for players to maintain close verbal contact if not physical contact and also work to bridge the gaps in physical space.

However, there is no unified measure of player enjoyment or even involvement in this type of digital gaming. Players are quite diverse in their responses about what they like about playing. Omar, for example, plays *Counter-Strike*, "because it's an activity that my friends and I do together for leisure. I generally do not play computer games when I am not playing against/with my friends. If the graphics were not as good as they are, then we would not play it. . . . I would not play it if I my friends were not involved."

Furthermore, most of the players we interviewed will only play shooter games if they are multiplayer and not single player. And the principle reason is the social interaction that such games provide. Omar commented, "I value the competition with others, whether friends or on the Internet. But the most important thing is that I can get a chance to do something with friends, which is free and enjoyable at the same time." For Lewis, "trading with other people . . . that's a huge reason why I play," referred to his enjoyment of *Diablo II*, a role playing game, but the same reason could apply to social interaction between players in *Counter-Strike*. When Omar played he signed on to an outside server and played with whoever was around. But, as he continued to play, other players became familiar and new social relations were established. Omar mentioned, "Over time, I started to get to know some people. . . . Like, honestly, I think it has brought me closer, normally we just talk online and stuff." A major reason why multiplayer games are preferred is the assumption that computer avatars are ultimately predictable and therefore boring, while human players are consistently doing something surprising or interesting.

On the other hand, Bruce, one of the older graduate students, who was more reserved than many of the other players, expressed his anxiety about revealing himself to strangers over the Internet. As he put it,

> If I don't know who I am playing with it feels very strange . . . because, I view it as a social activity . . . where I'm sort of represented in some sense by my character. . . . So, whenever I play on line with people I don't know I feel like I'm an outsider that's joining a group. . . . I had a microphone hooked up to myself yesterday for the first time. And there was someone from outside who had joined. And I noticed myself thinking at one point that, that I probably wouldn't use this microphone because this stranger is here. And I am sort of a reserved person anyway. So, it just felt like, it just felt strange. . . . I think, in a normal social situation I wouldn't sort of be very gregarious and outgoing. That sort of carries over into the way that I play and the way I view the whole thing.

Team playing or organized social interactions leading to a common outcome are a major reinforcement for *Counter-Strike* pleasure. Teamwork requires trust between players to pull off effective tactical moves. How that trust is established should be of interest to all game researchers. According to Frank,

> If you play on a server a lot and there's certain people that always come on the server, you get to know them, sort of, become, I guess you could say, quasi-friends, where, or acquaintances, where you know, you know the person, sometimes you're playing in groups and stuff like that, and you'd be on the same team, and I guess there's sort of a little bond there, you know, sometimes you'll go out together, on a, to kill someone, or you'll stick together in teams and stuff like that, but, it's all in good fun.

On the other hand Frank also made the point that social interaction between players is not the only reason for playing and that at certain times he just wants, "to go and, you know, seclude myself from the world for a little bit, and kick some ass I guess. So, I play it alone on the Internet." Like all social interaction, playing alone may be viewed as the flipside of playing together with friends. Both are necessary for game pleasure, although at different times. The other important aspect of trust between players is how to interpret insults. According to Frank, "playing with friends, sometimes it's a little more fun because I know them, I can poke fun at them, or they can poke fun at me." This "poking fun" is thought to heighten competition between players while also making it easier to joke with people who you think understand your intentions.

Of course, playing on a team is not just about coordinating your actions with friends; it also means having a relative game balance in abilities between players. The concept of equal skill level is significant in the pleasure that game players take away from their interactions. According to Daryl:

> I like when things are fair. It's almost, it's more enjoyable when, on one level, it's enjoyable when you know, the team I'm on, when I'm doing really well. It's also enjoyable when it's really back and forth. When there seems like a real, when it seems like an actual contest. Rather than you know, it's really not that much fun to bowl over somebody ten times in a row . . . it's boring.

In fact, Daryl complained about the different skill levels between the undergraduate players in the lab and his own team in the library, "You want, you want an appropriate level of difficulty. You know, when we played with some of those outside people, you know when we used to play [in the] library and the guys in the computer center. That was no fun, you know, they were just too good." Having equal skill levels facilitates closer relationships since resentment and frustration do not build from uneven game play.

Since these players were friends outside the game, game play served to draw them closer together. Jose mentioned,

> One thing is that during the game we are . . . kind of like closer and . . . then we kind of feel like we're on the same team. . . . Your relationship becomes closer when you are playing the game, but outside of the game it is our friendships you know, they extend to what the game, to the game playing and to interactions about that. We talk about what happened in the game and all that stuff and so we do have a relationship that we have in common. . . . Different types of people come together you know to relate in this game and then they go on to their lives, it's sort of like that movie . . . *Fight Club*.

Jose's understanding of this kind of play evoked a specific gender bonding, a safe place for men to get together and play. He said,

Male bonding . . . that's the way I look at it . . . got closer with the game. . . . I feel like I was bonding with my friends in this game, were mainly males because you know we don't have any females that come to play with us, but um I actually one day said you know why don't we go to hunting or something like that and actually we do these things for real because I went hunting once and I thought it was pretty exciting . . . one of them actually has been hunting with his friend Micky, and he had really been into this environment of weapons and all that. But all of the other ones, they're not, I don't think they saw it as fun, you know going and hunting, but they did see the area of male bonding and also just having fun and being able to you know, maybe have a couple of beers and just relating and talking about stuff so I think that yeah, our relationships were closer as a result of the game.

Player appreciation increases to the degree that team players can watch each other's backs and effectively coordinate their actions. As Jose mentioned, "someone tries to . . . kill you and you see . . . that your friend . . . got that other person that was gonna get you and so you appreciate that." While this bonding does make the players feel closer to their friends, it can also be viewed as an opportunity for time away from family and other friends. This was mentioned by several players. What may be understood by some friends as social isolation may, in fact, be just another move to a different set of friends and a different set of interests. Tim mentioned that he thought playing by yourself without friends was a problem, "It's kind of like the stigma of drinking alone."

The bar analogy was brought up several times by the game players in reference to a setting that induces social interaction. David said, "Yeah, it's fun to be on the same team, it's fun to be on a team with people you're nearby or people you can communicate with. Um, but it doesn't, I don't think it replaces social activity. . . . It's less social than being in a bar." On the other hand Tony, one of the undergraduate players in the lab, compared playing *Counter-Strike* to going to a bar, "Yeah, definitely, it's like going to the bar, just a different method. It's not conforming to whatever society has set to, this is what you do for fun, so I guess it's a different kind of hanging out I want to say." Daryl makes the point that if he had to stay home and not see his friends, not go to a bar, that would be a problem. But, he viewed playing *Counter-Strike* with those same friends in a different light. He said,

That is no big deal [not going to a bar with friends] because most of my friends are here. . . . I wouldn't, if I had to be home and shirked off going out with them . . . feel right but since we've got that ideal situation where we can all play in the same room, um, I mean that makes it, you know, pretty much the same as going to a bar because we still talk in between, you know, in between games and . . . our conversations are mundane enough at a bar that they, you know, can be repeated anywhere.

In our observations extensive talking, helping team players, pointing out where other players were moving, and other talk were recorded in our observations. At one point, one of the players even bought a fifth of Jack Daniels whiskey into the back library office, which they all laughed about and proceeded to drink.

Tina, who had played with a different group of game players, described playing *Counter-Strike* as expanding the way friendships were expressed. She said,

> If they're angry with each other or if they're frustrated with things in their relationship it gives them sort of a way to put that off. Cause, I've never left playing *Counter-Strike* feeling mad really. . . . I went there (to a Internet cafe) recently with one of my roommates and I was going to teach her to play *Counter-Strike* because she'd never played before and, like, her and I were having a really big, like, you know, relationship difficulty problem. . . . And when we went and I was teaching her, that sort of all went away, and it was just, you know, fun to play with her. . . . But we didn't actually solve anything and it was still there the next day but something about it kind of put that off and we were able to interact again and which was kind of nice, you know.

Tina's response raised the question of what purpose game playing serves in mediating relationships. The standard response that play is an escape from the world seems a bit too pat and simplistic, since it does not address the complexity of what escape actually means for the players involved and how that response is affected by where the game is played.

WHAT DOES IT MEAN TO "ESCAPE"?

As Tina's comments point out, an escape may be only a temporary pause while time passes and one waits, gathering resources to continue what had previously been brewing in the "real" world. Lewis said that playing *Counter-Strike*,

> that's my virtual world, if you will. When I go there, I'm somebody else, uh . . . I'm, uh, it's my world and it just completely blocks out all outside, anything." He equated playing *Counter-Strike* to sleeping, "if I want to stop thinking about someone I play *Counter-Strike*. Okay, I don't want to think about he or she anymore, and . . . to get, get away from the real world and its problems, I go into *Counter-Strike*. . . . An escape, yes. . . . It's like going to sleep I guess, like depressed people go to sleep sometimes . . . so I think playing *Counter-Strike* is like sleeping, you tend to forget about it for a while and then deal with it later, let time heal.

For Jose, playing *Counter-Strike* was stressful, but a stress which made him focus on the game and not his personal issues. He mentioned,

> Um, I think it is an exciting game and I think it is stressful, but I think it is stressful to a point where it is relaxing so I don't know if that makes any sense, you know, but I do feel . . . afterward I am relaxed and um, I think that *Counter-Strike* also gives me the ability to do something outside of my normal schedule and so it allows me to get out of that routine of everyday life. . . . That is what I value most. I get away from everything, and I do something that is only related to nothing. And I get to bond with the people that are there. I get to bond with the people that are there and I just get away from, from school, from friends, from any other problems that I have, you know, and it gets me, it gets all of my concentration. I don't think about anything else but the game and so that's why the game is good, because it gets you, it gets your full attention.

This desire to pursue play activity, as a "Liminal space" outside of one's normal routine, as an escape, works to sharpen and define the boundaries of everyday life and shape one's emotional responses towards significant events.[12] And yet such play space is fully integrated into their everyday lives, the often made rigid distinctions between work and play, are in fact, increasingly dissolving. The problem of using concepts like the self-contained play of the "magic circle" is eroded as players bring their game into real life stories and their work conversations into the game.

On the other hand, David had a more practical purpose for playing which was to take a break from grading papers, "After having finished grading sixty papers on a Wednesday afternoon and going to play at night, I was relieved." For those in a different line of employment (or non-employment), playing *Counter-Strike* will convey a different definition of escape. Tina, who was teaching English in Taiwan when she played the game, mentioned the pressures of teaching children, doing bilingual teaching, "I mean, it's really stressful and so, um, usually before I go home at night I stop for an hour, you know, and try to get some gaming in. I'll go home for a while and do some other stuff, whatever, catch up on stuff, grade their papers, do whatever, and then come back and play again later at night." Both play times and play spaces are therefore directly tied up in what one actually does and where one does it.

For the graduate students who were doing intense intellectual work during the day, a three hour gaming session with friends during the week was a welcome relief from constant academic pressure. As Tim put it, "For me . . . it's something to do with my friends . . . because it's such a social game. . . . Like I said there is some element of, 'I don't want to work on my dissertation now. I'm tired of thinking all day.' So, it's not so much family versus friends. It's more academic work versus an enjoyable time with friends." So while there is an obvious push to escape from routine work, school pressures, and family responsibilities, there is also a pull towards increased friendship solidarity and peer or affinity group support. For the graduate student players I interviewed, the very topics they had chosen for dissertations, the types of material they were studying evoked very powerful emotions of fear and anxiety in real life. According to Markus,

> I study philosophy and the philosophy I study is Contemporary Jewish Thought, which is in the wake of a great deal of violence. And so it's nice to have virtual stuff where you can still be testosterone laden and so forth and the electrons don't care, you know? After you reboot the computer, it's all good, nobody gets hurt, everybody has fun, rock on. Yeah . . . I study really deep, dark shit and it's nice to just not think about that sort of stuff for a while, um and it's fun.

Given that all the students, both undergraduates and graduates, had very busy schedules, the establishment of a three hour gaming session every Wednesday evening provided a vehicle for them to spend time with each other in a life situation where contact with one's friends was already highly attenuated. Bryson commented,

> It's certainly a nice pressure release. . . .You come on Wednesday night, run around, be crazy for awhile. But, you know, um, it would be a situation where given the busyness of everybody, I normally wouldn't see those guys unless we came together to play *Counter-Strike*. So, it sort of serves both functions, right? Takes me out of one facet of my life and puts me in another facet of it.

While the game allows one to temporarily suspend the concerns of off-line responsibilities, it also reinforces, through the playing of the game, new relationships and contacts. Clint mentioned, "I feel that it's brought me closer to my friends in the group cause in a sense we are doing a common activity even though we're in different parts of the office area. But out of that, I have developed various contacts. I've gotten together with the guys outside of that. So, it has in that regard, and it's also been a way just to get out of everything . . . that involves work or school and just go in with the guys and have fun."

DISCUSSION AND CONCLUSIONS

Given the above accounts, what can we learn about how the social spaces of play are produced and what they produce? And furthermore, what areas of interest do such conversations point to as fruitful places of research for social scientists? It appears to be common knowledge now that college students and their younger brothers and sisters have learned to multi-task, switching easily from one media form to another, integrating all of their media without leaving out any particular media. Rather than spending time in media isolation, for example simply watching television, today's college students, and indeed, youth in general, are intensely social. Digital games, and media in general, merely serve to facilitate this development. In a study in 2005 of 2032 young people between the ages of 8 and 18, the Henry J. Kaiser Family Foundation discovered that teenagers and pre-teens spent on average 6 hours a day in media consumption with 49 of those minutes spent playing video games.[13] Reading comprised 43 minutes of time. The rest of the time young people devoted to listening to music and watching TV. Media consumption, including playing video games, did not preclude social interactions, but quite the opposite, they encouraged more interaction and less isolation.

What is interesting about this data, and reflects upon our game player accounts, is that those children who have the most access to media were also those who were the most social, spending more time with friends, parents, and relatives, as well as more engaged in sports compared to the general population. Contrary to popular fears about the loss of reading skills and a decline in student literacy, in fact, "Media multi-taskers . . . those who spend the most time playing console video games (the 13% who play more than an hour a day) spend more time reading than those who play fewer video games."[14] This certainly appeared to be the case in our small sample. All the students were doing well academically, as well as being involved in either extracurricular activities or working on advanced research projects.

With the integration of media usage and the increasing sophistication of game narratives, game social relationships, and fan clubs, it appears that three major areas of interest are emerging, both from our data and from the literature. The first area is the rapidly fading Victorian era distinctions between play and work, between what is understood as legitimate versus illegitimate play, and the struggles that have emerged from the anxiety generated by this fading distinction. Thinking of game playing as merely a trivial "escape" illuminates not only a moral concern, but also a productive concern—the distinctions between games as a "waste of time" versus games as "teaching lessons, moral or otherwise" (serious, or productive games).[15] The word "escape" always seems to preface a need or desire to remove oneself from the everyday physical world of material necessity. In fact, in late modern capitalist culture, work and most of everyday life is heavily rationalized and play that is not just as rationalized (or professionalized) into producing surplus value or wealth is looked upon with suspicion, as "wasting time" or as not "serious."

The second area of interest is the relationship between game playing, social relationships, and place that appear throughout the data. Where one plays is important in establishing social relationships, but it is not set in stone. While playing from one's office or house can lead to making new online friends, it is when friends play together in the same spatial setting that the pleasure of game play is heightened. However, the mistake many critics make is to assume that face-to-face interaction is the only authentic or real interaction possible compared to the supposed impoverished interactions within cyberspace. I would hope that assumption has been put to rest. As our data demonstrates and as other studies have pointed out, this relationship is far more complex, where online and off-line worlds are interconnected more than many would like to admit. And why should we think differently? After all, telephones, cell phones, PDAs, TVs, computers, and the panoply of media use which characterize modern industrial social life facilitates interactions between people in ways impossible to imagine 50 years ago. Hence, both the private spaces managed with new digital technology as well as public spaces that are made comfortable for relaxed interactions increase, not decrease the capacity for people to come together. The increasing use of social network technologies like *Facebook* and *Skype* systems reveal a society bound closer together more than ever, with all of its inherent contradictions, tensions, and eccentricities. One sign of this is that many high status people make a point of not being in touch, as opposed to lower status workers who are always expected to be at the beck and call of their bosses.

Internet cafes can also work to increase social solidarity when designed properly to facilitate interaction. The question which emerges is not if people are connected or not, not if they are alienated or not, but rather what is the quality of that interaction? Game play, by bringing players together in a mutually pleasing activity inherently works to reinforce social solidarity by highlighting the pleasures of virtual and physical exchanges. But such game play always needs to be understood in the context of the real life social pressures players have to cope with in a crisis-ridden world. [16]

However, the question of the quality of social relationships mediated by the new technologies of representation raises a third point which is, who is excluded and who is included in those interactions? The idea of a digital divide between the haves and have nots is insufficient. Rather the divides can appear internal to those relationships in all media. For example, given the character of many FPS games, understanding the role that policing gender expectations has on social inclusions/exclusion is crucial for looking at what type of social community is facilitated and what type is discouraged. While it is obvious that males play FPS games more often than females, there is nothing inherently masculine about playing *Counter-Strike*. Rather the social construction of gendered social behavior by the dominant group in society has ascribed particular characteristics to types of behavior which then serve the purpose of regulating the behavior of that subordinated group. The fact that some women play *Counter-Strike* and enjoy other shooter games, by forming clans and clubs raises the issue of how we construct masculinity as violent and femininity as non-violent.

To conclude the new technologies of representation in interactive digital games and on-line virtual worlds amplifies already existing social relationships between players and non-players alike. Therefore to assume that such technologies in the form of digital games simply cause one to change their behavior in a negative direction or conversely to liberate one's fantasy or creative potential is to ignore the complex manner in which play spaces and everyday social relationships interact, establishing new understandings of old knowledge as well as reproducing well established meanings about the way the world works.[17] Digital games, therefore, are neither a panacea for a new utopian future nor should they be feared as violent messengers of an apocalyptic future. Such musings already circulate rapidly in modern society without the assistance of video games. But, they do raise the question of what role does fantasy and play serve in a hyper-capitalist world torn between the desire for growth and profit and the need for limited sustainable societies.

NOTES

1. This manuscript was presented in an earlier version at the Gamers in Society, Play in Culture Conference, The Third Annual Game Studies Seminar, University of Tampere Game Research Lab, Tampere, Finland, April 17–18, 2007.

2. Steve Jones, "Let the Games Begin: Gaming Technology and Entertainment Among College Students" (Pew Internet and American Life Project, 2003), accessed at http://www.pewinternet.org/~/media//Files/Reports/2003/PIP_College_Gaming_Reporta.pdf.pdf.

3. See for example, Constance A. Steinkuehler, "Massively Multiplayer Online Video Gaming as Participation in a Discourse," *Mind, Culture, & Activity*, *13*, 1(2006): 38–52; T.L. Taylor, *Play Between Worlds: Exploring Online Game Culture* (Cambridge, MA: MIT Press, 2006); and Dimitri Williams, "Virtual Cultivation: Online Worlds, Offline Perceptions," *Journal of Communication*, *56*, 1(2006): 69–87.

4. I am using the term affinity group as defined by Gee, "a group that is bonded primarily through shared endeavors, goals, and practices and not shared race, gender, nation, ethnicity, or culture." In James Paul Gee, *What Video Games Have to Teach Us about Learning and Literacy* (New York: Palgrave Macmillan, 2003), 197.

5. See Richard A. Bartle, *Designing Virtual Worlds* (New York: New Riders, 2004), 474.

6. See Henry Jenkins, "Game Design as Narrative Architecture," in *The Game Design Reader: A Rules of Play Anthology*, ed. by Katie Salen and Eric Zimmerman (Cambridge, MA: MIT Press, 2006), 670–689; and Jasper Juul, *Half-Real: Video Games between Real Rules and Fictional Worlds* (Cambridge, MA: MIT Press, 2005).

7. See the works of: Christopher Alexander, Sara Ishikawa, Murray Silverstein, with Max Jacobson, Ingrid Flksdahl-King, Shlomo Angel, *A Pattern Language: Towns, Buildings, Construction* (New York: Oxford University Press, 1977); Ray Oldenburg, *The Great Good Place: Cafes, Coffee Shops, Bookstores, Bars, Hair Salons, and Other Hangouts at the Heart of a Community* (New York: Marlowe and Company, 1999); and William Whyte, *The Social Life of Small Urban Spaces* (New York: Project for Public Spaces, Inc., 2001).

8. Rebekah Nathan, *My Freshman Year: What a Professor Learned by Becoming a Student* (Ithaca, NY: Cornell University Press, 2005).

9. William M. McDonald, *Creating Campus Community: In Search of Ernest Boyer's Legacy* (San Francisco, CA: Jossey-Bass, 2002).

10. Stephen Kline and Avery Arlidge, *Online Gaming as Emergent Social Media: A Survey.* (Burnaby, Canada: Media Lab, Simon Fraiser University, 2003). Accessed at http://www.sfu.ca/media-lab/onlinegaming/report.htm.

11. Kline and Arlidge, *Online Gaming as Emergent Social Media*, 1.

12. See the work of Victor W. Turner, *The Ritual Process* (Chicago: Aldine, 1969); Victor W. Turner, *From Ritual to Theatre: The Human Seriousness of Play* (New York: PAJ, 1982).

13. Victoria Rideout, Donald F. Roberts, and Ulla G. Foehr, *Generation M: Media in the Lives of 8–18 year-olds.* Executive Summary, Kaiser Family Foundation Study. Henry J. Kaiser Family Foundation (Menlo Park, CA, 2005). Accessed at http://www.kff.org.

14. Rideout, Roberts, and Foehr, *Generation M*, 2.

15. See Sonia Livingstone, *Young People and New Media* (Thousand Oaks, CA: Sage, 2002), 98. She clearly sees this manner of talking in the predominance of both the "money" metaphor versus the "diet" metaphor about media consumption; the money metaphor implies that one makes choices about the good and bad uses of one's time, to be productive, while the "diet" metaphor of media consumption assumes that one has to regulate their consumption to "cut down" on unproductive media usage. She says, "Interestingly, these metaphors for media use, centering on money and diet, share the ambivalence of an advanced industrialized society towards its own wealth: the culture values a media-rich, information-rich, nutritionally-rich, financially-rich society, but not one which is over-indulgent, lacking in challenge or 'fibre,' and not one which create the poverty—information-poor, media-poor, nutritionally-poor, etc.—against which such riches are distinguished."

16. As Furlong and Cartmel point out the expansion of individualization with the onset of Late Modernity, accompanied by neo-liberal capitalist policies of intensive exploitation of labor, have meant that social risk is increasingly negotiated by young people on an individual level accompanied by a decline in collective support, especially for those on the low end of the economic spectrum. This puts even more pressure on young people to connect with their peers via consumption to gain the support they are lacking in other areas of social life. See Andy Furlong and Fred Cartmel, *Young People and Social Change: Individualization and Risk in Late Modernity* (Philadelphia, PA: Open University Press, 1997).

17. Those who argue that media and computer use in general is causing the "death of childhood" are just as much mistaken as those who view computer technology and media as liberating children (Buckingham 2000; Holloway and Valentine 2003). Education, self-regulation, and the promotion of playing rights can move us beyond an empty moral debate to one which fully examines how all media, including game playing, is situated in our lives. See the work of David Buckingham, *After the Death of Childhood: Growing Up in the Age of Electronic Media* (London: Polity Press, 2000); Sarah L. Holloway and Gill Valentine, *Cyberkids: Children in the Information Age* (New York: Routledge, 2003).

Part II

Social Inequalities in Video Game Spaces: Race, Gender, and Virtual Play

Chapter Five

Racism in Video Gaming

Connecting Extremist and Mainstream Expressions of White Supremacy

Jessie Daniels and Nick LaLone

Millions of people play video games. In 2009, according to the National Public Diary Group (NPD Group), there were an estimated 169.9 million people playing video games in the United States.[1] Video games routinely make headline news due to their content, often for violence or for their supposedly addictive qualities. This growing visibility of gaming in the public sphere has led to a noticeable rise in video game studies which grow in complexity each year. However, what few engaged with video games acknowledge—whether playing, designing, reporting on, or analyzing games—is the presence of racist content. During the early days of the Internet, some scholars theorized that the emergence of virtual environments and a culture of fantasy would mean a rise in *identity tourism*,[2] that is, people using the playful possibilities of gaming to visit different racial and gender identities online. However, the reality that has emerged is quite different. The rise of the popular Internet has shown that racial and gender identities offline are transported, relatively untransformed, into digital constructs, such as video games.[3] Emerging research suggests that players vigorously enforce conformity to offline gender identity, as well as gender norms, in online gaming.[4] What has been left largely unexamined until now are the complex ways in which *systemic racism*,[5] both overt and subtle, is implicated in video gaming culture. Neither the scholarly literature on race nor the research on video gaming has taken up the challenge of exploring the intersection of racism in gaming.

In this chapter, we begin the work of addressing this gap in the literature by first examining the crude video games created or co-opted by members of white supremacist movements such as *Border Patrol*. Then, we turn to more popular games, such as *Grand Theft Auto III & IV* and *Saints Row 1 & 2*, and explore the more subtle racism in these games. We place both kinds of games within the dual context of a prevailing Internet culture in which humor is often the most highly valued commodity, and a wider social context that is supposedly post-racial and largely dismisses charges of racism. By contrasting the various ways systemic racism is both displayed and enacted in these disparate video games, we illuminate the connections between extremist and more mainstream forms of white supremacy.

BACKGROUND: WHAT DOES GAMING TEACH US?

According to Huizinga, the way we play is the way we understand the world.[6] This is most obviously expressed through the way we play games and is uniquely represented through video gaming because video games represent how imaginary and real systems work.[7] These ways of constructing and playing in different worlds can be an important mechanism of socialization into the offline world for people who play them. This is sometimes referred to as "the real" world of materiality.[8] People learn the implicit rules of society through the explicit rules of play.[9] Because those same people create video games, each designer's view of social reality provides a means through which cultural practices are communicated. For instance, video games that rely on explicit rules of competition and mastery in order to "be number one" among a field of opponents upon penalty of (virtual) death convey important lessons about what it takes to survive in an (actual) economic system premised on a neoliberal ideology of individual striving and a vanishing safety net.

Everything in video games—opening a door, detonating a nuclear bomb, breathing, driving a car, shooting a gun, falling from outer-space, or interacting within bureaucratic government systems—must be painstakingly programmed. Video game makers can represent almost any object or action possible in society, as well as many that are impossible. The interaction between a video game's design and the player's choices in the game allows players to form and act on opinions of how those systems work and the parameters that confine actions within each system. Video game researcher Bogost refers to this interaction as *procedural rhetoric.*[10] This procedural rhetoric is far from neutral; in many games, it is laden with systemic racism.

Far from being disembodied and "race-less," the information age is as racialized as the previous industrial age.[11] Cultural studies scholar Lisa Nakamura criticizes this notion of the Internet as a "race-less" utopia in her book *Cybertypes*. In that book, she demonstrates precisely how, through interface design elements like pull-down menus with categorical lists of racial and ethnic identities, the online world reproduces racial identity constructed offline. This idea that racial oppression is linked to visibility is one that African American scholars have written eloquently about going back to W.E.B. Du Bois. This idea also appears in the literature on "race" and the Internet. Some scholars have argued that the Internet offers a freedom from the visibility of racial oppression through the "decoupling identity from any analogical relation to the visible body."[12] Yet, the supposed invisibility online rests in part on the assumption of the Internet as an exclusively text-based medium in which racial identity is not visible. While that may have been true at one point in time, or may be true today in certain online contexts, it does not adequately describe much of what constitutes life online these days.

Today, the Internet includes digital video and photographic technologies, such as "webcams" along with photo-sharing sites like Flickr.com and video-sharing sites like YouTube.com. Most social networking sites, such as MySpace.com and Facebook.com, prominently feature visual elements such as digital photos and sometimes video that serve as important markers of digital representation and identity for people who participate at those sites (indeed, as the name "Facebook" suggests the notion of linking visual representation of the physical body to text is embedded in the purpose of the software). These inherently visual technologies make images of bodies a quotidian part of the gendered and racialized online world. Furthermore, empirical research increasingly demonstrates that people go online, even to text-only online spaces, not as a libertarian utopia of disembodiment, but as a mechanism for engaging in the construction and affirmation of embodied racial identities and these identities are in turn, shaped by power relations.[13] To the extent that race is discussed in the scholarly literature about the Internet, it is usually framed around issues of racial and ethnic identity in online communities.[14] Little of this scholarship has discussed racism in video gaming.

Who Is Playing?

Video games are overwhelmingly made by and for males.[15] They are the playful embodiment of what is commonly referred to as Human-Computer Interaction (HCI). Justine Cassell, in writing about genderizing HCI,[16] observes that the formative years of technology use occur as early as kindergarten and that this training shapes, and continues to shape, how males and females perceive technology. Boys are encouraged to explore the parameters of computer technology while girls are encouraged to use computer technology as simple tools. Thus, HCI is an important form of anticipatory socialization that provides a gateway for boys to learn more about how computer systems work and contributes to the continued male dominance of most technology-based industries.[17] Research indicates that girls and women report less frequent video game play, less motivation to play games in social situations, and less orientation to game genres featuring competition and three-dimensional rotation.[18] This reluctance may be attributable evidence that demonstrates gaming culture is hegemonically white, heterosexual, and masculine. For example, Lori Kendall argues in her richly nuanced ethnography of the gendered dynamics in the multi-user domain (MUD) *"BlueSky,"* that digital technologies reproduce white, heterosexual, masculine cultures and hierarchies of power.[19] While this is by no means conclusive evidence, the preliminary research does indicate a pattern of white male dominance of online spaces.

Video games are conceptualized as similar to, or at the very least coterminous with, Internet technologies. Like Internet technology, video games are often thought to be a "white" activity because of how media presents them.[20] American-created video games have reflected the racial environment of post–World War II American technological development through the make-up of mostly white male game programmers as well as the procedures or systems included in the games themselves.

Video gaming is predominantly created by white people and played most often by white people. According to the Pew Internet and American Life Project, a random sample of 12–17 year old children in the United States found that 73.9 percent of all white children play video games while 26.1 percent of all non-white children play (See table 5.1). Many video game researchers contend that games are made mostly by white males, however, it is difficult to find demographic data specifically about game makers. The best data available is related to technology jobs more generally. Recent data shows that more than 90 percent of all Silicon Valley job markets are held by whites.[21] Gender patterns in high-tech jobs and entrepreneurship have historically reflected traditional patterns of gendered job segmentation, with women in low-paying, limited advancement positions (such as assembly work) and men in higher-paying, career-ladder jobs (such as game designers).[22] These patterns have begun to shift slightly in the last decade, with more women entering non-traditional sectors especially as entrepreneurs, yet the research finds that women entrepreneurs tend to not own businesses in "male-typed" high-tech sectors, such as game design firms.[23] Thus, there is ample evidence to suggest that game design, like the high-tech industry as a whole, is a white and male-dominated industry. What this means is that from user interface design to hardware design, it is predominantly white males who design, test, and distribute video games. This is done while the importance of video games as a major vehicle of socialization in human computer interaction increases.

Video games are both a recreational activity and an educational tool, but it is as recreation that video games are marketed. As with any recreational activity, those with more income and leisure time are more likely to purchase and play video games. However, reliable data on who gamers are is often muddled by operationalization of the term video game. For example, Nielson reports that, "the most active gamers tend to be younger males in the 12 to 17 range, living in homes that have incomes of $75,000 or more."[24] This study separates video games between PC and console markets but then uses combined (PC and console) numbers to report on female game use of casual games (e.g., *Solitare* or *Hearts*). Inclusion of casual games typically leads to higher estimates of the number of players with higher percentages of women and a greater diversity of ethnic groups. In comparison, the NPD Group released the *Essential Facts About the Computer and Video Game Industry.*[25] This study reported that as of 2009, the average age of video gamers is 34 years of age but that "women age 18 or older represent a significantly greater portion of the game playing population (33%) than boys 17 or younger (20%)." However, NPD Group combines the PC and console markets. Further, this study, like many widely cited studies of gamers, is funded primarily by the game industry which has a vested interest in inflating these numbers.[26] ·

Table 5.1. Video Game Play amongst Children

Q4			Race		
			White	Non-White	Total
Does your son or daughter ever play video games?	Yes	719	254	973	
			73.9%	26.1%	100.0%
	No	91	21	112	
			81.3%	18.8%	100.0%
Total		810	275	1085	
			74.7%	25.3%	100.0%

"BORDER PATROL": GAMES CREATED BY MOVEMENT WHITE SUPREMACISTS

Tom Metzger—a former Ku Klux Klan (KKK) leader, a television repairman by trade, and a one-time candidate for Congress—combines the elements of both showcasing and private uses of the Internet at his website, "The Insurgent." Metzger's web presence, like his former print media incarnation called "W.A.R.," an acronym for "White Aryan Resistance," is a showcase for white supremacist ideology in which white, heterosexual men are central, people of color are referred to as "mud people," and Jewish people are thought to control banking and media in an international conspiracy to keep down the white race.[27] Metzger left the KKK in 1983 when he formed W.A.R. and developed a more radical analysis of political economy than the KKK and dropped any reference to Christianity.[28] To spread the message of W.A.R., Metzger created both print and broadcast vehicles: a newsletter, titled "W.A.R.," a cable access television show called "Race and Reason," and a radio broadcast.[29] All these media are now showcased and available via Metzger's website, "The Insurgent" located at the URL www.resist.com. The website includes position statements about a variety of topics including immigration, international conflicts (most often involving Israel), homosexuality, and women. Prominently featured on the website is a link to purchase Aryan-branded merchandise (t-shirts, caps, key chains). The merchandise page includes the use of some forms that require a user login but to actually place an order, the end-user has to print out and mail in an order form with a check or money order. Aside from these forms, most of the features on the website are primarily static, and function as one-way transfers of information.

One of the noteworthy features on Metzger's website, because it is unique to the digital media environment and was not available during the print-era, is the selection of hate-filled computer games. These games, with names like *Drive By 2* where players can experience *What it is Like in the Ghetto*, *African Detroit Cop, Watch Out Behind You Hunter*, situate gamers as shooters (in the convention of video games). Interestingly, most of these games originate on humor sites like *Newgrounds.com* and are downloaded and re-branded by Metzger as games for and by white supremacists. In these racist games, players are instructed to "shoot the fags before they rape you." In the game called *Border Patrol*, with the tag line, "Don't Let Those Spics Cross Our Border," gamers are encouraged to "shoot the spics." The games allow individual users to download and play the games on their own computers. In addition to being violently racist and homophobic, the computer games are also deeply gendered in ways that are consistent with more mainstream games; that is, the games socialize boys into misogyny and exclude girls from all but the most stereotypical roles.[30] Research clearly demonstrates that adolescents are more likely to play computer games than adults; among adolescents, boys are more likely than girls to be gamers.[31] Adolescents are also significantly more likely than adults to say that violence is their favorite part of gaming.[32] Metzger has included these computer games on his website to appeal to his core audience: young, white males. However, without an evaluation of his internal website statistics, which are not publicly available, it is impossible to know how effective Metzger's racist games are with the intended audience.

Metzger's computer games are crude bits of gaming code that barely adhere to standards in gaming[33] and seem unlikely to meet the minimum demands of sophisticated gamers who have grown up playing increasingly sophisticated games. While games like the ones that show up on Metger's site are considered crude, these games do present an opportunity for players to explore or act upon beliefs that may be otherwise hidden from the public. *Border Patrol* made headlines in 2009 when a Georgia councilman was forced to resign after emailing a link to the game to employees of Kennesaw, Georgia's local government.[34] Conversely, the higher production values of popular video games use the typical color-blind rhetoric as a building block for the procedures represented in their systems. The use of this rhetoric creates opportunities for players to perform what would normally be private acts, at home, alone, or with friends.

GRAND THEFT AUTO & *SAINTS ROW*: SUBTLE RACISM IN POPULAR GAMES

Some of the most complicated procedural rhetoric created for mainstream consumption can be found in the *Grand Theft Auto* series. These video games represent thousands of hours of development time from hundreds of people working to represent a particular cross-section of society. We are examining two installments of the *Grand Theft Auto* series: *Grand Theft Auto III: San Andreas* (*GTA: SA* and *Grand Theft Auto IV* (*GTAIV*). These two games represent the opposite coasts through their settings (Los Angeles and New York City) as well as two different perspectives of minority groups. *GTAIV* represents an immigrant's path to respect and honest work in the harsh, racially segregated, urban environment.

Grand Theft Auto: San Andreas is about an African American male named Carl "CJ" Johnson. Like Metzger's games, but perhaps with less intention, *Rockstar* has embedded certain African American racial stereo-types within the game. First, nearly every non-white character is represented as a gang banger. While some characters approach CJ from outside this realm

(James Woods provides the voice of a "government agent"), they almost always treat him as the ontologically suspect black man from the inner city. [35] The similarities between CJ Johnson and Eddie Murphy's character in the game *African Detroit Cop* are rather astonishing in the way that they both replay centuries-old tropes of systemic racism. Both games display African Americans as hapless, violence-causing miscreants that will only change their anti-social behavior if threatened with loss of money. In *GTA: SA*, racial stereotypes are conveyed via food choices. Throughout the game, African American characters are restricted to food choices from a limited range of fast food restaurants with names meant to evoke the ghetto. Even when CJ owns a large portion of California at the end of the game, he can still only eat at three different fast food restaurants in the game: Burger Shot, Cluckin' Bell, and The Well Stacked Pizza. CJ gains and loses weight according to the number of meals he eats per day. The implications of the programming behind this meal plan for the African American protagonist is that CJ's blackness is inherent and immutable; the upward mobility possible by succeeding in the game does not offer an escape from this embodiment. In the rhetoric of the game, CJ will always be black and therefore, will always favor fast food. [36] This digital representation of blackness has little to do with actual material experience of any individual black person, rather the digital represents the game programmers' misperception of the embodiment of blackness.

In contrast to CJ Johnson is Niko Bellic, the white immigrant from Eastern Europe, who is the central figure in *Grand Theft Auto IV*. Niko comes to America after events that occurred during the Bosnian War come back to haunt him. His life is one of violence and he will commit violence for almost anyone. From steroid-addled white body builders to Puerto Rican female drug dealers, Niko will commit violence without remorse on anyone he is asked to. This game approaches racism toward immigrants from European nations that have not become part of the "white" group. The stereotypes that *GTAIV* picks up on are the same ones that were used in the movie *2012* as well as *Borat: Cultural Learnings of America for Make Benefit Glorious Nation of Kazakhstan.* However, unlike CJ who remains part of the inner city forever, Niko inevitably obtains his place in white society in a dramatic fashion by murdering the corrupt white foreign businessman at the foot of the statue of liberty. While CJ is doomed to stay in the inner city, Niko, a newcomer, bypasses all of them. History has shown that this acceptance cannot be asked for, only accessed through hard work. [37]

One of *Grand Theft Auto's* direct competitors is the gangland simulation, *Saints Row*. It differs from the *Grand Theft Auto* series by presenting a much bleaker, more violent picture of an urban gang environment. From the introduction, the player is thrust into a world of violence and bloodshed. The solution, as stated by the leader of the "3rd Street Saints," says, "[I]t's all about respect. Get enough of it, they're gonna back off and we're gonna move right on in."[38] This introduction serves two distinct racially motivated ideas. First, the solution is not hard work to gain power, but respect. Respect comes through violence and fear of that violence is how one gains respect. This reflects a general fear of minority groups gaining power and is distinct in a variety of Metzger's videos and video games. This fear comes to a head at the end of the game when a corrupt white businessman tells the main character, "[U]ntil you came along, I was displacing poor people. Now I'm destroying a hotbed of gang activity." The last scene of *Saint's Row* is of that businessman's yacht exploding.[39] Unlike in Metzger's games, overt racism is punished in this world.

Saint's Row 2 creates an environment in which the urban gangs gain power in popular society alongside government entities and enforcement agencies. However, from the beginning mission it is clear that the successes of the player's character in the previous game have been corrupted. The first impulse of the main character is to kill all of the last game's characters and kill their way back to the top. There is no discernable goal other than to "gain respect." At the end of this game, the character who began *Saint's Row* saying, *"It's all about respect"* is revealed as an ally of the white man. This character says, as he dies, *"Don't you get it? The Saints didn't solve a goddamn thing . . . all we did is turn into vice kings that wore purple."*[40] The *Saint's Row* games almost directly reference the undercurrent of systemic racism running through the dominant white culture in the U.S. by designing a situation in which the player can only succeed by committing violence and perpetuating racist stereotypes.

These four games are a primarily white interpretation of African American culture for white people to play. In this way, video games represent a way in which systemic racism is expressed differently than in the offline world. The creation of a private sphere to explore the thoughts and values game makers put into their games allows players to explore certain aspects of their racially motivated beliefs that they may not know or understand they have. Many of the currents that run through these games are not labeled as racist until they end up on a website like Metzger's. And, because Metzger's site is poorly designed and because his games are "crude," many players brush those games off as "hate group messages." After getting off the computer, that person then proceeds to go to a virtual inner city and "do" racism as Metzger intends. In this way, players can do racism, privately, in the backstage[41] without any sort of repercussions for their actions.

POST-RACIAL: RACISM IN THE ERA OF COLOR BLINDNESS

Racism, both overt and more subtle, in video games exists in a social context in the contemporary United States that is at odds with such displays of racism. The prevailing view in the United States is that some fifty years after the civil rights era and with the election of an African American president, the highest ideal and most appropriate moral response to racism is one of color blindness, or "not noticing" race.[42] The ideological orientation toward color blindness has implications for racism online. A recent study links color-blind racial ideology to racism online and off.[43] The study examined the relationship between responses to racial theme party images on social networking sites and a color-blind racial ideology, and found that white students and those who rated highly in color-blind racial attitudes were more likely not to be offended by images from racially themed parties. In other words, the more "color-blind" someone was, the less likely they would be to find parties at which attendees dressed and acted as caricatures of racial stereotypes (e.g., photos of students dressed in blackface make-up attending a "gangsta party" to celebrate Martin Luther King Jr. Day) offensive. To conduct the study, Tynes and Markoe showed 217 ethnically diverse college students images from racially themed parties and prompted them to respond as if they were writing on a friend's Facebook or MySpace page. Fifty-eight percent of African-Americans were unequivocally bothered by the images, compared to only 21 percent of whites. The majority of white respondents (41 percent) were in the bothered-ambivalent group, and 24 percent were in the not bothered-ambivalent group. In the written response portion of the study, the responses ranged from approval and nonchalance ("OMG!! I can't believe you guys would think of that!!! Horrible . . . but kinda funny not gonna lie") to mild outrage ("This is obscenely offensive"). The participants were also asked questions about their attitudes toward racial privilege, institutional discrimination, and racial issues. Those who scored higher on the measure were more likely to hold color-blind racial attitudes, and were more likely to be ambivalent or not bothered by the race party photos. Respondents that scored low in racial color blindness were much more vocal in expressing their displeasure and opposition to these images, and would even go so far as to "de-friend" someone over posting those images. Yet, that culture of supposed "colorblindness" is set within a broader social reality of systemic racism that includes the historical racism of slavery, contemporary manifestations of institutional discrimination, and ongoing individual acts of racism.

FUN AND GAMES: TAKING RACISM SERIOUSLY

Mainstream Internet culture is centrally concerned with humor. For evidence of this, one need only refer to the immense popularity of LOL cats, an Internet meme and a booming online subculture built around digital images and deliberately bad grammar. A few months after launching the site icanhascheezburger.com, the site receives around 200,000 unique visitors, a half-million page views each day, and ad revenues earn the site's owner a comfortable income.[44] Even the attack on the World Trade Center is fodder for Internet jokes through visual collages, assembled from phrases and pictures from popular media.[45] The high value placed on humor has implications for understanding responses to racism.

Within this *milieu*, it can be difficult to challenge racism or take it seriously. Take, for example, the practice of "racist griefing" in online video games. "Griefing" in online gaming is similar to "trash talking" to opponents that might happen on a basketball court or a football field; in gaming, griefing happens in online interactions, and oftentimes it becomes explicitly racist (e.g., opponents typing "NIGGER, NIGGER, NIGGER" at one another). Lisa Nakamura makes the point that the "racist griefing" that goes on in online games often makes explicit use of racist epithets, which she explains this way: "The n-word is funny because it is so extreme that no one could really mean it. And humor is all about 'not meaning it.' If you take humor and the n-word, you get enlightened racism online and attention." Nakamura goes on to argue that paradoxically, "the worse the racism and sexism are, the more extreme and cartoonish it is, the harder it is to take seriously, and the harder it is to call it out." She astutely observes that for those within gaming culture, calling out racism in this context signals you as someone "not of the gaming culture" and thus, as someone who is taking racism "too seriously" and does not have a good sense of humor. Yet, this sort of humor is a "confusing discursive mode for young people," she observes, because they are "unable to separate enlightened racism from regular racism." And, indeed, I think this is a real problem here. As Nakamura notes, the image of the "humorless feminist" is now joined with the image of a "humorless" old(er) person who takes race too seriously.[46] Within gaming culture, humor trumps any concerns about the harms of racism.

A similar phenomenon is evident in the exuberant embrace of the comedy of Leeroy Jenkins from the game *World of Warcraft*, which has become a hugely popular Internet meme. Leeroy Jenkins is a character originated by Ben Schulz (who is white). The encounter within the game starts as a group of friends are attempting to defeat a monster. During a conversation between the leaders of the group, Jamaal and Abduhl talk about strategy and tactics to defeat the monster that sits on the other side of a door. In the midst of discussing their strategy, Leeroy Jenkins (one of the group who has been marked as "AFK" or away from the keyboard, supposedly to microwave some chicken). The groups' plan is ruined when Leeroy returns and, ignorant of the strategy, charges headlong into battle shouting his own name in a stylized battle cry. His companions rush to help, but Leeroy's actions ruin the meticulous plan, and all of the group members are killed. At the end of this battle, while his friends deride him for his reckless behavior, Leeroy can be heard saying, "Well at least I have chicken." This interaction within the game has been captured on YouTube video which now has over 23 millions views. The popularity of the Leeroy Jenkins meme reaches well beyond the *World of Warcraft* players, and into the broader popular culture[47] and many of these are racialized. For example, the site "You're the Man Now Dog" (YTMND), includes a picture of an African American man in medieval armor (Martin Lawrence from his movie *Black Knight*) standing in front of a Kentucky Fried Chicken and a bucket of chicken repeating the phrase, "at least I have chicken" to the song "I Got the Power."[48]

Leeroy Jenkins is one of the only representations of blackness in *World of Warcraft*, a universe in which all humans are white.[49] The humor in the Leeroy Jenkins meme is dependent on a number of racist stereotypes about "hapless negroes" who are so impulsive and distracted by their love of chicken that they miss what's important and destroy those around them (and themselves). Yet, any attempts to take the racism in this humor seriously get dismissed by gamers. As Tanner Higgin observes about these failed attempts at calling out the racism in the Leeroy Jenkins humor, "these questions are buried beneath claims of comedy and the insignificance of race in the game world."[50] While Higgin contends that racism in online gaming is easy to dismiss because it is ephemeral, unlike racism in the material world, gaming scholar Christopher Ritter disagrees. For Ritter, the minimization of racism in online games is an extension of minimization of racism elsewhere.[51] In this way, confronting the systemic racism in gaming culture is rendered impossible through both the denial of racism in "color blindness" and the valuation of humor above all else.

CONCLUSION

Racism exists in online games in a variety of forms, both overt and more subtle. Even as video games increase in popularity, few within gaming culture acknowledge the systemic racism in many of these games. Here, we have argued that the reasons for this are multifaceted. Simultaneous with vitriolic racist hate speech, often spread via the Internet and video games, the dominant white culture claims to be "color-blind" and dismisses concerns about racism as irrelevant. Added to this is an Internet culture, also predominantly white, in which humor is the highest value and charges of racism are regarded as the purview of the humorless and the overly serious. The racism in online video games is built into the very procedural rhetoric of the games, yet remarking upon race is seen as more problematic than the harm of racism. And pointing out racism marks one as an outsider to gaming culture.

Predictions of the Internet's early days imply an escape from the material realities of race and expansion of identity tourism. Video game design and play provide designers and gamers, primarily white males, *entrée* to and sinecure in a hegemonic space within the worlds of the digital and the material. In some ways, the contradictory and overlapping qualities of overt racism (e.g., in Metzger's games) and the more subtle racism in popular video games speaks to the paradoxical nature of racism that characterizes the current historical moment.

NOTES

1. Tor Thorsen, "US gamer population: 170 million—NPD." (GameSpot, July 31, 2009), accessed at http://www.gamespot.com/news/6214598.html.
2. Lisa Nakamura, *Cybertypes* (New York: Routledge, 2002).
3. Lisa Nakamura, *Digitizing Race* (New York: Routledge, 2008); Dara N. Byrne, "The Future of (the) 'Race': Identity, Discourse, and the Rise of Computer-Mediated Public Spheres." In *Learning Race and Ethnicity: Youth and Digital Media*, ed. by Anna Everett (Cambridge, MA: MIT Press, 2008), 15–38.
4. Zeke Valkyrie, "Gender in MMORPG's," paper presented at the annual meeting for the American Sociological Association, Atlanta, Georgia, August 2010.
5. In this chapter, we use the term *systemic racism* to refer to the way that enduring and systemic racial stereotypes, ideas, images, emotions, proclivities, and practices have thoroughly pervaded social, cultural, and economic institutions. This includes both subtle and overt expressions. Significant changes have occurred in systemic racism over time, primarily through the political struggles organized by people of color, yet it remains a central feature in most major social institutions. For more on this view, see Joe R. Feagin, *Systemic Racism: A Theory of Oppression* (New York: Routledge, 2006).
6. John Huizinga, *Homo Ludens* (Boston, MA: Beacon Press, 1950).
7. Ian Bogost, *Persuasive Games: The Expressive Power of Videogames* (Cambridge, MA: MIT Press, 2007).
8. P.M. Greenfield, "Video Games as Cultural Artifacts," In *Interacting with Video*, ed. by P.M. Greenfield and R.R. Cocking (Norwood, NJ: Ablex, 1996), 85–94; P.M. Greenfield, L. Camaioni, P. Ercolani, and L. Weiss, "Cognitive Socialization by Computer Games in Two Cultures: Inductive Discovery or Mastery of an Iconic Code?" *Journal of Applied Developmental Psychology, 15*, 1(1994): 59–85.
9. Karen E. Dill & Kathryn P. Thill, "Video Game Characters and the Socialization of Gender Roles: Young People's Perceptions Mirror Sexist Media Depictions," *Sex Roles 57* (2007): 851–864.
10. Bogost, *Persuasive Games.*
11. Jessie Daniels, *Cyber Racism: White Supremacy Online and the New Attack on Civil Rights* (Lanham, MD: Rowman & Littlefield, 2009).
12. Mark B.N Hansen, *Bodies in Code: Interfaces with Digital Media* (New York: Routledge, 2006), 145.

13. Byrne, "The Future of (the) 'Race': Identity, Discourse, and the Rise of Computer-Mediated Public Spheres."

14. See, for example, Emily Ignacio, *Building Diaspora: Filipino Community Formation on the Internet* (New Brunswick, NJ: Rutgers University Press, 2005) and Matthew Hughey, "Virtual (Br)others and (Re)sisters: Authentic Black Fraternity and Sorority Identity on the Internet," *Journal of Contemporary Ethnography 37*, 5(2008): 528–560.

15. Brenda Laurel, *Utopian Entrepreneur* (Cambridge, MA: MIT Press, 2001); Yasmin Kafai, Carrie Heeter, Jill Denner, and Jennifer Y. Sun, *Beyond Barbie and Mortal Kombat: New Perspectives on Gender and Gaming* (Cambridge, MA: MIT Press, 2008).

16. J. Cassell, "Genderizing HCI." In *The Handbook of Human-Computer Interaction,* ed. by J. Jacko and A. Sears (Mahwah, NJ: Lawrence Erlbaum, 2002), 402–411.

17. Cassell, "Genderizing HCI"; Judy Wajcman, "From Women and Technology to Gendered Technoscience," *Information, Communication and Society 10*, 3(2007): 287–298.

18. Kristen Lucas and John L. Sherry, "Sex Differences in Video Game Play: A Communication-Based Explanation," *Communication Research 31*, 5(2004): 499–523.

19. Lori Kendall, *Hanging Out in the Virtual Pub*: *Masculinities and Relationships Online* (Berkeley: University of California Press, 2002).

20. Dmitri Williams, Nicole Martins, Mia Consalvo, and James D. Ivory, "The Virtual Census: Representations of Gender, Race and Age in Video Games," *New Media Society 11* (2009): 815–834.

21. Mike Swift, "Five Silicon Valley Companies Fought Release of Employment Data, and Won," *San Jose Mercury News,* February 14, 2010, accessed at http://www.siliconvalley.com/news/%20ci_14382477.

22. Susan S. Green, "Silicon Valley's Women Workers: A Theoretical Analysis of Sex-Segregation in the Electronics Industry." In *Women, Men and the International Division of Labor,* ed. by June Nash and Maria Patricia Fernandez-Kelly (Albany: SUNY Press, 1983), 273–331.

23. Heike Mayer, "Segmentation and Segregation Patterns of Women-Owned High-Tech Firms in Four Metropolitan Regions in the United States," *Regional Studies 42*, 10(2008): 1357–1383.

24. Gavin McMillan, "The State of the Video Gamer PC Game and Video Game Console Usage Fourth Quarter 2008." (The Nielsen Company, 2009), accessed at http://blog.nielsen.com/nielsenwire/wp-content/uploads/2009/04/stateofvgamer_040609_fnl1.pdf.

25. "The NPD Group," *Demographic and Usage Data: Essential Facts About the Computer and Video Game Industry* (National Purchase Diary, 2010), available for download at http://www.npd.com/.

26. "The NPD Group," *Demographic and Usage Data.* "The NPD Group," *Average Age of Gamer* (National Purchase Diary, 2010), accessed at http://www.npd.com/press/releases/press_100527b.html; "Pew Research Center," *February 2008—Teen Gaming and Civic Engagement* (Pew Internet & American Life Project, 2008), available for download at http://pewinternet.org/Shared-Content/Data-Sets/2008/February-2008--Teen-Gaming-and-Civic-Engagement.aspx; Gavin McMillan, "The State of the Video Gamer PC Game and Video Game Console Usage Fourth Quarter 2008."

27. Jessie Daniels, *White Lies* (New York: Routledge, 1997); Daniels, *Cyber Racism.*

28. For an in depth analysis of Metzger, there are a number of good resources, most notably see: Rafael Ezekiel, *The Racist Mind: Portraits of American Neo-Nazis and Klansmen* (New York: Penguin, 1996) and Eleanor Langer, *A Hundred Little Hitlers: The Death of a Black Man, the Trial of a White Racist, and the Rise of the Neo-Nazi Movement in America* (New York: Picador, 2004). For a broad overview of many of the groups discussed here, including more about Metzger, see Kathy Marks, *Faces of Right Wing Extremism* (Boston: Branden Books, 1996).

29. Morris Dees and Steve Fiffer, *Hate on Trial: The Case Against America's Most Dangerous Neo-Nazi* (New York: Villard Books, 1993); Wallace Turner, "Extremist Finds Cable TV Is Forum For Right-Wing Views," *New York Times*, October 7, 1986: A23.

30. Justine Cassell and Henry Jenkins, *From Barbie to Mortal Kombat: Gender and Computer Games* (Cambridge, MA: MIT Press, 2000).

31. Cassell and Jenkins, *From Barbie to Mortal Kombat.*

32. M.D. Griffiths, N.O. Mark, and Darren Chappell, "Online Computer Gaming: A Comparison of Adolescent and Adult Gamers," *Journal of Adolescence 27*, (2004): 87–96.

33. Katie Salen and Eric Zimmerman, *Rules of Play: Game Design Fundamentals* (Cambridge, MA: MIT Press, 2004).

34. Luke Plunkett, "Councilman Resigns After Spruiking Racist Flash Game," *Kotaku* March 19, 2009, accessed at http://kotaku.com/5174556/councilman-resigns-after-spruiking-racist-flash-game.

35. "Wikipedia," *List of* Grand Theft Auto: San Andreas *Characters,* (Wikipedia, N.d.), accessed at http://en.wikipedia.org/wiki/List_of_characters_in_Grand_Theft_Auto:_San_Andreas#Michael_Toreno.

36. Bogost, *Persuasive Games.*

37. See U.S. Supreme Court decision *Ozawa v. United States*, 260 U.S. 178, (1922), accessed at http://supreme.justia.com/us/260/178/case.html.

38. BonersGames, "Saint's Row: Intro and First Mission: Canonized" (YouTube.com, April 24, 2007), accessed at http://www.youtube.com/watch?v=pXb5wArKJCs.

39. BrujoX, "100% Saints Row Ending" (YouTube.com, October 26, 2006), accessed at http://www.youtube.com/watch?v=FpBe4L0Lr5A&feature=related.

40. raphu64, "Saint's Row 2: Secret Mission (Revenge on Julius) [SPOILER]" (YouTube.com, October 21, 2008), accessed at http://www.youtube.com/watch?v=5rqjmzG2yEo&feature=related.

41. Leslie Houts Picca and Joe R. Feagin, *Two-Faced Racism: Whites in the Backstage and Frontstage* (New York: Routledge, 2007).

42. Eduardo Bonilla-Silva, *Racism without Racists: Color-Blind Racism and the Persistence of Racial Inequality in the United States* (Lanham, MD: Rowman & Littlefield, 2003/2006).

43. Brendesha M. Tynes and Suzanne L. Markoe, "The Role of Color-Blind Racial Attitudes in Reactions to Racial Discrimination on Social Network Sites," *Journal of Diversity in Higher Education 3*, 1(2010): 1–13.

44. Aaron Rutkoff, "With 'LOLcats' Internet Fad, Anyone Can Get In on the Joke," *Wall Street Journal*, August 25, 2007, accessed at http://online.wsj.com/article/SB118798557326508182.html.

45. Giselinde Kuipers, "Media culture and Internet disaster jokes: bin Laden and the attack on the World Trade Center," *European Journal of Cultural Studies 5*, 4(2002): 450–470.

46. Lisa Nakamura, "Don't Hate the Player, Hate the Game: Internet Games, Social Inequality, and Racist Talk as Griefing," presentation at the Berkman Center for Internet & Society, Harvard University, June 16, 2010, accessed at http://blogs.law.harvard.edu/mediaberkman/2010/06/16/lisa-nakamura-dont-hate-the-player-hate-the-game/.

47. For example, the television game show *Jeopardy* featured a question about Leeroy Jenkins, accessed at http://www.urbandictionary.com/define.php?term=Leeroy Jenkins.

48. Hammerson, "You're the Man Now Dog: Let's Do This! Leroy Jenkins!" accessed at http://leroyjenkins.ytmnd.com/.

49. Christopher Jonas Ritter, "Why the Humans are White: Fantasy, Modernity, and the Rhetorics of Racism in *World of Warcraft*" (PhD diss., Washington State University, 2010).

50. Tanner Higgin, "Blackless Fantasy: Disappearance of Race in Massively Multiplayer Online Role-Playing Games," *Games and Culture 4*, 1(2009): 3–26.

51. Ritter, "Why the Humans are White: Fantasy, Modernity, and the Rhetorics of Racism in *World of Warcraft*," p.120.

Chapter Six

Worlds of Whiteness

Race and Character Creation in Online Games

David Dietrich

Video games are an enormous part of popular media today, comparable to television and movies. In 2007, revenue from video games amounted to 9.5 billion dollars,[1] only slightly less than the 9.63 billion dollar revenue of U.S. movie theatres.[2] Moreover, like Hollywood movies, video games are very much in the mainstream of American society, despite popular misconceptions of games as entertainment for male adolescents.[3] Sixty-five percent of American households play computer or video games; forty percent of game players are women; and the average game player today is 35 years old.[4] Moreover, video games represent a new form of media, distinguished from previous forms due to the interactive element where game players have the ability to change and influence the game world.[5] Online games, particularly Massively Multiplayer Online Role-Playing Games (MMORPGs), have created an entirely new form of communication, where people interact with each other in a computer-generated virtual world.[6] Players are represented in this world by "avatars," computer-generated characters molded by the players within the limitations of a particular game. These avatars serve as the sole representation and means of interaction for the player within the game world.

Social scientists are just beginning to examine many of the issues associated with this new form of media. Although we have a large body of literature addressing the effects of popular media on perceptions of race,[7] comparable research on the effects of racial representation in video games is in its infancy. This paper will contribute to the study of race and popular media by examining how race is presented in online video games (MMORPGs) through a particular feature: avatar creation. Specifically, I

101

document how many online games restrict or disallow the creation of avatars with a non-white racial appearance. This has the potential to shape the social composition of the online world by creating all-white virtual spaces, which reinforces a sense of normative whiteness within the game world.[8] This has potential consequences for racialized interactions within the virtual space[9] as well as outside the game.[10]

ANALYSIS OF RACE IN THE MEDIA

There is a large body of literature that addresses media representation of race. Generally, the focus of the existing literature is on the perpetuation of racial stereotypes. Minority characters in entertainment media are routinely circumscribed in how they are portrayed, leading to disproportionate stereotypical representation.[11] For example, media tend to portray blacks as subservient or sexually violent or deviant (e.g., pimps, prostitutes), Native Americans as savage warriors, Latinos as criminals, and Asian Americans as asexual "helpers" or "nerds."[12] Scholars and journalists have also noted the lack of minority representation in the news media,[13] as well as the general exclusion of minority groups, including blacks and Latinos, from entertainment media.[14] While offering valuable data, these media studies of race have been limited to traditional forms of media, including movies,[15] television,[16] and print media, such as newspapers and magazines.[17] There is a dearth of scientific study of interactive media, such as the Internet and electronic games.

Within the emerging field of video game studies, most of the studies of minority representation have focused on issues of gender.[18] Only a handful of studies have addressed the issue of stereotypical portrayal of race in games. Furthermore, social scientists, particularly sociologists, have neglected this area compared to scholars from the humanities. Leonard makes a plea for the importance of race-based research on games, pointing out not only the stereotypical portrayal of minority characters in games, particularly blacks, but also the lack of non-white player-characters in games.[19]

Among studies of stereotypical racial representation in games, Jahn-Sudman and Stockmann described the portrayal of blacks, particularly black men, as violent criminals in the game *Grand Theft Auto: San Andreas.*[20] DeVane and Squire also examine this game, this time looking at how players' own experiences and knowledge lead to differing interpretations of racial stereotypes within the game.[21] In a similar vein, Blackmon and Terrell examined how players perceive race in the game *Grand Theft Auto: Vice City.*[22] Additional studies have focused on the representation of Asians in video games. Sisler focused specifically on stereotypical representation of

Arabs and Muslims within a framework of terrorism and hostility. [23] Thomas charted the creation of a player group within *Diablo II* called KPK, Inc. (Korean Player Killers) who were dedicated to seeking out and virtually "killing" characters played by Korean players. [24] The KPK player group was formed in response to perceived threats to American game servers from Korean players who engage in "farming," [25] real-money trading, [26] and lag [27] caused by Korean players overloading American servers. Using a different approach, Everett examined how racial stereotypes are presented through user manuals and strategy guides for games, using both visual and textual analysis. [28] Finally, a recent study by Eastwick and Gardner found differences in reactions of players to requests based upon the skin color of the requesting avatar, with dark-skinned characters receiving less favorable responses. [29] This suggests that the racial appearance of in-game avatars has a real social effect among players in the game space.

Studies that document the absence of non-white players in games include Higgin, [30] who examined blackness in the context of MMORPGs. Higgin noted how black player-controlled characters are not only largely absent from games in general, but also how race, when not portrayed in stereotypical terms, is ignored in terms of any effect within the game world. In another recent study, Williams, Martins, Consalvo, and Ivory conducted a large-scale examination of race, gender, and age representation among characters in video games (both primary and secondary) in 150 of the best-selling games across nine platforms in 2005. [31] They found a systematic underrepresentation of females, racial minorities (excluding Asians), children, and the elderly.

RELEVANT SOCIOLOGICAL THEORY

I argue that the lack of minority player-controlled characters in video games, particularly in online games, reinforces ideas of normative whiteness and has the potential effect of reinforcing existing ideologies of white supremacy in the United States and elsewhere.

Research in the area of whiteness studies demonstrates how whiteness is rendered "default" or "normal" within the context of contemporary racial ideology. [32] That is, whiteness becomes the category against which difference is measured. In this context, whiteness itself loses any categorization as a "race." Rather, it simply is. As Garner states, "treating whiteness as a non-racialized identity conceals racialized power relations and the ideas and practices that sustain them." [33] Whiteness becomes a universalizing category, encompassing all the privileges granted to whites by virtue of skin color and masking issues of how white privilege contributes to non-white disadvantage. Furthermore, as a universalizing category, whites are viewed as individuals while non-whites are viewed as parts of groups.[34] This serves to deflect charges of group-based blame or undue privilege for whites while at the same time engendering stereotypical, group-level interpretations of acts committed by racial minorities.[35] These group-level stereotypes lead to the interpretation that it is in the nature of racial minorities to commit deviant acts, while whites that do so are "aberrant," acting on individual impulses.[36] With regard to issues of racism, this interpretation eliminates all group-level responsibility for whites, and indeed whites often respond with denial, declaring racism to result from individual prejudices.[37]

While most existing research has focused on media perpetuation of negative stereotypes about minorities, the same holds true for perpetuating normative whiteness. Harwood and Anderson argue that media representations not only reflect the conditions of society but maintain them as well.[38] The lack of minorities in the media reinforces the view of whiteness not as a race, but as "the" race.[39] From the view of normative whiteness, seeing white people everywhere is normal and natural because white people are not white, they are simply people. At the same time, the stereotypical and circumscribed presentation of minorities reinforces the idea of the "Other," where "people" (whites) interact with racial minorities who are driven by aspects of their collective "nature." The process through which such ideas concerning non-whites are formed is termed "white habitus"[40] and occurs as a result of social and spatial segregation of whites. That is, due to the separation of whites from non-whites in society, there emerges a sense of white "groupthink" that rationalizes racial differences, where whites create positive self-views that justify their advantaged racial position and negative views of racial "others" to explain racial inequality.[41] Thus, whiteness becomes the normative condition, white advantage the standard against which others are measured, and inequality the result of purported negative qualities of racial minorities.

Within interactive media, such as games, such presentations take on an additional dimension: when one plays a game, it becomes more than simply observing but also participating. That is, the player is not only interacting *with* the media, but *within* the media.[42] The player, in a sense, exists in two places at once: sitting in front of a screen, hands on keyboard or controller, while simultaneously within the game world embodying an avatar, a computer-generated character controlled by the player that has the (limited) power to change and affect that world.[43] This situation becomes even more complicated when we think about interactions made in multiplayer online games. The other avatars in the game world are not merely semi-autonomous constructs of pixels and code but graphical representations of *other* players, occupying the same spaces. In this case, the player character, or avatar, has the potential to become more than just a role taken on by the player but a virtual representation of the actual player. As such, most MMORPGs give you the ability to customize your avatar, in order to allow the player to tailor his or her virtual representation. In this context, I argue, the lack of minority representation becomes even more problematic in that interactions between avatars (that is, between two individual players) become racialized virtually, due to the constraints of appearance and behavior imposed by the game itself on the player's avatar. This leads to the reproduction and reinforcement of existing racial attitudes and views by creating a virtual white habitus, where virtual characters exist within an all-white world due to limitations in creating non-white characters. Given this concern, I decided to examine the degree to which it is possible to create non-white avatars in online games, focusing on MMORPGs.

METHODOLOGY

MMORPGs are characterized by players interacting with one another in an online virtual space, with thousands of players able to interact with one another in real time through their player-controlled characters. I set out to conduct a comprehensive survey of the character creation capabilities of all MMORPGs operating and accepting U.S. players [44] in the spring and summer of 2009. Unfortunately, there is no authoritative listing of operating Massively Multiplayer Online Game (MMOs). Consequently, I constructed a list from three online sources: *Gamespot* (www.gamespot.com), a major online gaming website that provides news and reviews of games; *MMOGData* (mmogdata.voig.com), a website dedicated to providing subscription data for MMOs; and *OnRPG.com*, a website that advertises itself as "the biggest and best Free MMORPG Directory on the net." Using information from all three sites, I compiled an initial list of 220 actively operating MMORPGs, including both games that have monthly fees and those that allow play for free. The total number was reduced in the course of downloading or attempting to purchase play time for these games, as a number were either no longer in operation, had web sites or servers that were not operational, or had clients (executable game programs) that would not run.

Additionally, in order to determine the capabilities of online games for creating non-white player characters, I eliminated those games that did not have a visible human [45] avatar (for example, games where you only controlled a vehicle or games without graphical interfaces). I also eliminated those games where the racial characteristics of the avatar could not be determined, for example due to size/quality of the graphical presentation. Third, because the purpose of my research was to see what constraints were placed on players who attempted to create a non-white virtual representation, I focused only on those games that allowed character customization. After eliminating operating games that did not fit into the above criteria, I ended up with a total sample of 65 MMORPGs.

To determine the ability of a player to create a non-white avatar, it was necessary to specify measures of racial representation for these avatars. This was difficult due to the variations in graphical clarity and art style present among the games. Furthermore, in most cases it was not as simple as declaring the "race" of the avatar; one has to choose the individual aspects of the avatar's appearance, including hair style, color, face (sometimes individual features), skin color, clothing, height, and weight, among others. This creates the possibility of a player choosing, for example, a non-white skin color but being restricted to white-looking faces and hairstyles. Because of this, I decided to separately examine and record three readily visible racial markers: skin color, hair style/color, and facial features. Skin color and hairstyle were visible in all games, while facial features were identifiable in that subset of games that offered greater graphical detail. To objectively compare skin color between different games, I found it necessary to use a skin tone scale. Though somewhat problematic, I decided to use a variation of the Von Luschan chromatic scale to measure skin tone.[46] This scale provides gradations of skin tone from white to dark in 36 levels. The scale was later modified for use in identifying sun tanning risk,[47] collapsing the 36 levels into six categories from very light to very dark.[48] I attempted to create a character with the "darkest" skin tone available and recorded the color using the above categories to determine the maximum available deviation from "white" skin color. (See figure 6.1.)

Most games did not offer a level of graphical detail where hair texture could be determined. Therefore, I looked at hair color and style to see if any hairstyle typically associated with coarse (African) hair (e.g. afro, dreadlocks)[49] or any hairstyle capable of being worn by someone with coarse hair (no straight hair or loose curls) was available for selection and if the hair color selection allowed for sufficiently dark hair colors. Finally, determination of facial features (for those games with distinguishable facial features) was based upon examination of the size and shape of the nose and lips.[50]

I attempted to create a non-white-looking human avatar in each game, recording the darkest available skin tone using the six categories, the available black-associated hairstyles and darkest hair color, and the available facial options. The latter included determining if the faces could be chosen from a set of pre-made selections, or if the facial features themselves could be directly adjusted, and if so, the degree to which it was possible to make a non-white or African-looking face.[51] I focused on black or African features for two reasons. First, dark skin color offers one of the most visible and easily identifiable racial features in games with the most crude graphics, and there is a set of hairstyles specifically associated with coarse, African hair

	1	10			19	28	
	2	11			20	29	
	3	12			21	30	
	4	13			22	31	
	5	14			23	32	
	6	15			24	33	
	7	16			25	34	
	8	17			26	35	
	9	18			27	36	

Figure 6.1. Von Luschan's Chromatic Scale

that are easily identifiable. Similarly, facial differences between whites and blacks are easily noticeable in games with a minimum level of graphical detail. The same is true for Asian facial features as well. I note where Asian features were available, but the art style in many games is such that characters declared specifically to be Asian are indistinguishable from white presentations (an interesting feature that I will discuss more fully later).

RESULTS

Skin Color

Of the 65 games examined, 26 of them had no options to change skin color whatsoever and, in keeping with the findings regarding normative whiteness, the unchangeable, "default" skin color was white. (See table 6.1.) Those games that allowed for skin color changes did so in three different, but ultimately equivalent ways. Most allowed the player to choose skin color from a set of pre-generated options, while others allowed the player to make detailed adjustments using a color slider or color wheel, and still others explicitly connected player skin color to a preset "character" that comprised a predetermined set of skin, face, and hair options. Thirty-nine of the examined games allowed for some kind of alteration of skin color. However, of those 39, only 16 allowed you to choose skin colors of the darkest category (VI), while 18 others had selections in the category V range.

Three games (*Lord of the Rings Online*, *Dark Age of Camelot*, and *Final Fantasy XI*), while having skin color options, have such light tone choices that it would be difficult to distinguish between a non-white character and a character who was simply tanned. In the case of *Final Fantasy XI*, the difference is almost so subtle as not to be noticed, except that particular skin color is attached to a preset that includes a "cornrow" hairstyle and somewhat African facial features. This setting was the only non-white (and non-Asian) character preset available (out of eight). Another note of interest is that some games had fictional human "races" within the game world itself. For example, in *Age of Conan*, non-white skin colors were only available among characters of the Stygian race, who are described in-game as "dusky skinned." In some cases, these in-game races seem to draw upon existing racial stereotypes. Within *EVE Online*, not only is there only one race with non-white skin colors (Minmatar, a race of former slaves), but only one "bloodline" within that racial group has non-white skin colors (Brutors, a "swarthy" people that "favor physical prowess over anything else").

Hair Color/Style

Hair colors were generally selectable and ran the gamut from very dark to light, and in many cases included unnatural hair colors like blue and green. Hairstyles, on the other hand, were overwhelmingly variations on straight hair (with a handful of curly), but very few hairstyles that could be naturally worn by those with "African" hair, including something as simple as short-cropped natural hair. (See table 6.2.) Forty-two of the games examined had no such hairstyles. Furthermore, among those games that did, many only had one or two available compared to many more straight hairstyles. Of the 23 games that had African hairstyles, 12 had only one such option. Only six games had more than two African hairstyles to choose from. The "cornrow" style was the most common, followed by Afros, along with a few options for dreadlocks and simple short-cropped natural hair. Again, some games re-

Table 6.1. Skin Color Options

None	26 (40.0%)
Category VI	16 (24.6%)
Category V	18 (27.7%)
Category IV and below	5 (7.7%)
Total	65 (100.0%)

stricted these hairstyles to particular presets. For example, in *Vanguard: Saga of Heroes*, there were multiple types of humans, one of which was clearly black, which featured all African hairstyles. Similarly, *EVE Online* only had African hairstyles for the aforementioned Brutors.

Facial Features

As mentioned previously, some games could not be analyzed in terms of facial features due to the art style or lack of graphical detail necessary. (See table 6.3.) A total of 60 games had identifiable facial features, and of those, the ability to create non-Anglo faces, including wider noses and fuller lips, varied dramatically. Ten games had no ability to change facial features at all, and the default faces, of course, had Anglo features. Of those games that had the option to alter faces, there were two methods. Most games had a selection of pre-made faces that the player could choose from. A few, however, gave the player the ability to "morph" the face, allowing manual adjustment of things such as nose size and width, chin length, eye placement, and cheek-bones, to name a few of the options. The specific parts of the face that could be changed and the limits on those changes varied from game to game. Among those games that had facial selection or modification, most did not allow for non-white-looking faces. Of those games with selectable pre-made faces, 22 had no non-white faces. Nine games had identifiably non-white faces, while another nine had faces that were somewhat ambiguous.[52] Like hairstyles, there were often only one or two non-white faces among many other white ones. Six of the above nine only had one or two non-white faces among their selections. Only three games had multiple non-white faces from which to choose.

Though only ten games implemented facial morphing (or selection of individual facial features), one would assume this to be a solution to the above problem: if the player can change the face however he or she wishes, the player should be able to easily create a non-white face (though, as noted above, having hair styles or appropriate skin color to go along with it is another matter). However, the reality is that significant restrictions are placed on facial morphing features that make it difficult to make a non-white-look-

Table 6.2. "African" Hairstyle Options

None	42 (64.6)
One hairstyle	12 (18.5%)
Two hairstyles	5 (7.7%)
Three or more hairstyles	6 (9.2%)
Total	65 (100.0%)

Table 6.3. **Facial Feature Options**

Face Selection		
	White faces only	22 (36.7%)
	Non-white faces	9 (15.0%)
	Ambiguous	9 (15.0%)
Morphing/Feature Selection		
	Non-white faces impossible	5 (8.3%)
	Non-white faces impossible	5 (8.3%)
None		10 (16.7%)
Total		60 (100.0%)

ing face. Five games have morphing options sufficient to make a non-white face. Some only allow, for example, limited modification of the nose (*Luna Online, Dark Age of Camelot*) while others disallow sufficient lip changes (*Lord of the Rings Online*). One game, *Entropia Universe*, while having facial morphing options, does not allow the player to change the nose or lips to make an African-looking face. Another point of interest is in the game *EVE*, where, while having facial morphing features available for a character of any "race," only the Brutors have features that can be made into something resembling an African face.

DISCUSSION

The impetus for this research was my inability to make a convincing black character for the MMORPG City of Heroes due to lack of appearance options. When I began this research, I was familiar with a handful of MMORPGs, primarily those by big names that charged monthly fees (e.g., *World of Warcraft, Everquest, Age of Conan*). I was aware that these games were usually lacking in all the options needed to make a non-white character: many had skin color options, but few had appropriate facial features or hairstyles. However, I was surprised at the extent to which non-white appearance options are lacking among the broader population of MMORPGs. After collecting the names of every MMORPG I could find on the Internet and testing all those with customizable human characters, only *four* (*Dungeons and Dragons Online, Perfect World International, Star Wars Galaxies*, and *Vanguard: Saga of Heroes*) had the ability to create a "black" character. This number increases to eight if you include those games that allow a non-white skin tone but do not have options for the darkest skin tones (*Anarchy Online*,

Deicide, *EVE Online*, and *Football Superstars*). And of these games, two of them restrict black-looking characters to specific in-game races (*EVE Online* and *Vanguard: Saga of Heroes*).

The most obvious consequences of this situation are that the vast majority of virtual worlds out there are lacking truly non-white (specifically black) characters; they simply cannot be made in those environments. This includes games like *World of Warcraft*, which boasts around 10 million players worldwide, and *Second Life*, with over 12 million players.[53] It is one thing to live in a world where most white people live within a white habitus, sheltered from contact with racial and ethnic minorities, but it is quite another to explore, socialize, and play in a virtual world where one is not simply isolated from non-whites, *non-whites simply do not exist*. As more people throughout the world come online and virtual communities like *Second Life* continue to grow and new communities are created, the potential impact on the perceptions of race throughout the world is troubling. We know that popular culture is "both producer and product of social inequality,"[54] but the advent of MMORPGs adds another element beyond the passive traditional media, one whose impact is not yet fully understood: interactivity. How will people be influenced by interacting in online environments devoid of racial minorities? Will minorities be less likely to join these environments if unable to create avatars that represent themselves? These are but a few of the questions that the results of this study bring to mind.

To the contrary, there are potential arguments that say that this lack of racial options is not an oversight or omission, but merely a result of the game production process in terms of either the game's setting, its country of origin, or both. For instance, one could argue that a game like *World of Warcraft*, for example, is justified in its exclusion of non-white characters because the game is set in a sword-and-sorcery fantasy world derived from European medieval society. Other games, such as *Zu Online*, are similarly set in a fantasy version of medieval China, and thus non-Asian characters, in that case, would be out of place. However, such arguments fall short when one considers that these are *fantasy* worlds; there is no necessary connection between the historical places and periods from which they draw and the finished product. Indeed, *World of Warcraft* allows non-white skin colors for its human characters (but lacks hairstyle and facial options to go with them). While *Zu Online* has only Asian characters, other games set in similar Asian fantasy settings have non-Asian options (e.g., *Perfect World International*). These are ultimately fictional worlds, and the creators and maintainers of these worlds have the power to decide whom to let in as they wish.

A similar argument has to do with the origins and intended audiences of many of these games. Many MMORPGs, particularly free-to-play games, are produced in Asian countries like Korea and China, where the racial climate differs from that of the U.S. Many of the MMOs available in the United

States are simply localized versions of games that already have been running for years in parts of Asia (e.g., *Final Fantasy XI*, *Perfect World International*). One could argue, then, that the lack of non-white options in such games is understandable, as blacks do not form a significant portion of the population in these areas of the world. However, the fact is that regardless of where the game is made and the reasons for the lack of minority representation in games, the inability to create minority characters in these games has the same effect on American players as American games. Interestingly, it is somewhat telling that all these games made in countries like Korea can be so easily imported to the United States without any major graphical changes. If a lack of minority character options is due to lack of minorities in the countries of creation, then one might expect all the playable characters to have Asian features, yet most do not. The wide range of white-looking characters in these games made half a world away suggests much about the dominance of whiteness in world culture. [55]

CONCLUSION

This research complements and expands the recent work by Williams et al. [56] by addressing a specific gap they noted in their own research: that of racial representations in MMORPGs. At the same time, this research has some limitations of its own. Because good statistics for subscriber and player numbers of MMOs do not exist, there is no reliable way to determine which games may have a disproportionate effect by virtue of their player base. While *World of Warcraft* is widely known for its millions of players, no reliable numbers are available for the dozens of free-to-play games like *Maple Story*. Furthermore, the impact of these games must be examined in a global context, but there is no reliable way to know the geographic location of the various players of all these games. Finally, I examined race only among human characters, in part because trying to determine where green skin and tusks fit in among whites, blacks, Latinos, and Asians is not an easy task. However, a superficial examination of many games suggests that stereotypical portrayals of real-world minorities are being used to construct virtual-world races. For example, within *World of Warcraft*, both the Tauren (a race of bovine humanoids) and Orcs (a race of humanoids with green skin, tusks, and otherwise monstrous features) have fictional histories that seem to be drawn directly from the history and popular conceptions of Native Americans. An examination of how race might be playing out among non-human "races" in online fantasy games could be a fruitful avenue of exploration.

The size of these games and their respective virtual worlds continues to grow, as does the number of people interacting with each other through these mediums. But to date, many of these virtual worlds are not just dominated by white characters, the virtual worlds are *solely* white. We are living in the nascent era of online communities and collective virtual spaces, so it is premature to draw conclusions about their effects. Nevertheless, if we are to take heed of what we have learned from research on traditional media, we should be afraid. We are living in the age of what one "webcomic" has called the "World of Whitecraft."[57]

NOTES

1. Entertainment Software Association, "Industry Facts" (ESA Entertainment Software Association, 2011), accessed athttp://www.theesa.com/facts/index.asp.

2. Sue Ziedler, "U.S. Movie Box Office Hits Record in '07: MPAA," *Reuters*, March 5, 2008, accessed athttp://www.reuters.com/article/PBLSHG/idUSN0536665220080305.

3. Dmitri Williams, "A Brief Social History of Game Play." In *Playing Video Games: Motives, Responses, and Consequences*, ed. by P. Vorderer and J. Bryant (Mahwah, NJ: Lawrence Erlbaum, 2006), 229–247.

4. Entertainment Software Association, "Industry Facts."

5. Barry Atkins, "What Are We Really Looking At? The Future-Orientation of Video Game Play," *Games and Culture 1* (2006): 127–140; James Paul Gee, "Why Game Studies Now? Video Games: A New Art Form," *Games and Culture 1*, (2006): 58–61; Torben Grodal, "Video Games and the Pleasure of Control." In *Media Entertainment: The Psychology of Its Appeal*, ed. by D. Zillmann and P. Vorderer (Mahwah, NJ: Lawrence Erlbaum), 197–214; Peter Vorderer, Silvia Knobloch, and Holger Schramm, "Does Entertainment Suffer from Interactivity? The Impact of Watching an Interactive TV Movie on Viewer's Experience of Entertainment," *Media Psychology 3*, 4(2001): 343–363.

6. Elaine Chan and Peter Vorderer, "Massively Mulitplayer Online Games." In *Playing Video Games: Motives, Responses, and Consequences*, ed. by P. Vorderer and J. Bryant (Mahwah, NJ: Lawrence Erlbaum, 2006), 88–103; Tony Mannien, "Interaction Forms and Communicative Actions in Multiplayer Games." *Game Studies 3*, (2003), accessed at http://www.gamestudies.org/0301/manninen/.

7. See Tamara Baldwin and Henry Sessoms, "Race and Ethnicity." In *Media Bias: Finding It, Fixing It*, ed. by W. Sloan and J. Mackay (Jefferson, NC: McFarland, 2007), 105–116; Clint C. Wilson, Felix Gutierrez, and Lena M. Chao, *Racism, Sexism, and the Media: The Rise of Class Communication in Multicultural America* (Thousand Oaks, CA: Sage, 2003).

8. Steve Garner, *Whiteness: An Introduction* (New York: Routledge, 2007); Jane Ward, "White Normativity: The Cultural Dimensions of Whiteness in a Racially Diverse LGBT Organization," *Sociological Perspectives 51* (2008): 563–586; Ladelle McWhorter, "Where Do White People Come From? A Foucaultian Critique of Whiteness Studies," *Philosophy and Social Criticism 31* (2005): 533–556.

9. For example, Houts Picca and Feagin describe how interactions among whites in white-only "backstage" situations are characterized by more overt racist discourse that those in racially mixed company. See Leslie Houts Picca and Joe R. Feagin, *Two-Faced Racism: Whites in the Backstage and Frontstage* (New York: Routledge, 2007). If an MMORPG disallows the creation of non-white characters, there is the possibility of the entire game world becoming a "virtual backstage," where all interactions seem to be among whites, judging by the appearance of in-game avatars.

10. Stephanie Greco Larson, *Media & Minorities: The Politics of Race in News and Entertainment* (Lanham, MD: Rowman & Littlefield, 2006); Wilson et al., *Racism, Sexism, and the Media.*

11. Larson, *Media and Minorities*; Baldwin and Sessoms, *Race and Ethnicity.*

12. John Downing and Charles Husband, *Representing 'Race': Racisms, Ethnicities, and Media* (Thousand Oaks, CA: Sage, 2005); Larson, *Media and Minorities*; Wilson et al., *Racism, Sexism, and the Media*; Dana Mastro and Elizabeth Behm-Morawitz, "Latino Representation on Primetime Television," *Journalism and Mass Communication Quarterly 82* (2005): 110–130.

13. Larson, *Media and Minorities*; Baldwin and Sessoms, *Race and Ethnicity.*

14. John Gabriel, "'Dreaming of a White . . . '" In *Ethnic Minorities in the Media: Changing Cultural Boundaries*, ed. by S. Cottle (Philadelphia, PA: Open University Press, 2000), 67–82; Dan Collins, "Hispanics on TV: Barely a Cameo," *Associated Press*, June 25, 2003, accessed athttp://www.cbsnews.com/stories/2003/06/25/entertainment/main560313.shtml; Associated Press, "NAACP Calls for More Blacks on TV," December 18, 2008, accessed at http://newsone.blackplanet.com/entertainment/naacp-calls-for-more-blacks-on-tv; Baldwin and Sessoms, *Race and Ethnicity.*

15. Wendy Leo Moore and Jennifer L. Pierce, "Still Killing Mockingbirds: Narratives of Race and Innocence in Hollywood's Depiction of the White Messiah Lawyer," *Qualitative Sociology Review 3* (2007): 171–87; Norman K. Denzin, "Screening Race," *Cultural Studies, Critical Methodologies 3* (2003): 22–43.

16. Katrina Bell-Jordan, "Black, White, and a Survivor of *The Real World*: Constructions of Race on Reality TV," *Critical Studies in Media Communication 25* (2008): 353–372; Brian F. Schaffner and Mark Gadson, "Reinforcing Stereotypes? Race and Local Television News Coverage of Congress," *Social Science Quarterly 85* (2004): 604–623; Scott Coltrane and Melinda Messineo, "The Perpetuation of Subtle Prejudice: Race and Gender Imagery in 1990s Television Advertising," *Sex Roles 42* (2000): 363–389.

17. Rhea Sengupta, "Reading Representations of Black, East Asian, and White Women in Magazines for Adolescent Girls," *Sex Roles 54* (2006): 799–808; Melvin E. Thomas and Linda A. Treiber, "Race, Gender, and Status: A Content Analysis of Print Advertisements in Four Popular Magazines," *Sociological Spectrum 20* (2000): 357–371.

18. See Pamela Takayoshi, "Gender Matters: Literacy, Learning, and Gaming in One American Family." In *Gaming Lives in the Twenty-First Century*, ed. by S. Selfe and G. Hawisher (New York: Palgrave Macmillan, 2007), 229–249; Nowell Marshall, "Borders and Bodies in *City of Heroes*: (Re)imaging American Identity Post 9/11." In *Computer Games as a Sociocultural Phenomenon*, ed. by A. Jahn-Sudmann and R. Stockmann (New York: Palgrave Macmillan, 2008), 140–149; Jeroen Jansz and Raynel G. Martis, "The Lara Phenomenon: Powerful Female Characters in Video Games," *Sex Roles 56* (2007):141–148; Carrie Heeter, Rhonda Egidio, Punya Mishra, Brian Winn, and Jillian Winn, "Alien Games: Do Girls Prefer Games Designed by Girls?" *Games and Culture 4* (2009): 74–100; Rika Nakamura and Hanna Wirman, "Girlish Counter-Playing Tactics," *Game Studies 5* (2005), accessed at http://www.gamestudies.org/0501/nakamura_wirman/.

19. David J. Leonard, "Not a Hater, Just Keepin' It Real: The Importance of Race- and Gender-Based Game Studies," *Games and Culture 1* (2006): 83–88.

20. Andreas Jahn-Sudmann and Ralf Stockmann, "Anti-PC Games: Exploring Articulations of the Politically Incorrect in *GTA San Andreas*." In *Computer Games as a Sociocultural Phenomenon*, ed. by A. Jahn-Sudmann and R. Stockmann (New York: Palgrave Macmillan, 2008), 150–161.

21. Ben DeVane and Kurt D. Squire, "The Meaning of Race and Violence in Grand Theft Auto," *Games and Culture 3* (2008): 264–285.

22. Samantha Blackmon with Daniel J. Terrell, "Racing toward Representation: An Understanding of Racial Representation in Video Games." In *Gaming Lives in the Twenty-First Century*, ed. by C. Selfe and G. Hawisher (New York: Palgrave Macmillan, 2007), 203–216.

23. Vit Sisler, "Digital Arabs: Representation in video games," *European Journal of Cultural Studies 11* (2008): 203–219.

24. Douglas Thomas, "KPK, Inc.: Race, Nation, and Emergent Culture in Online Games." In *Learning Race and Ethnicity: Youth and Digital Media*, ed. by Anna Everett (Cambridge, MA: MIT Press, 2008), 155–174.

25. "Farming" refers to the repeated killing of a creature or creatures in order to harvest rare or valuable items dropped by those creature(s).

26. Real-money trading is the practice of either selling in-game items for actual money, or using money to purchase such items.

27. Lag refers to the time it takes for an action taken by the player at his or her computer to actually register within the game world. Lag can have many causes, such as network congestion, where commands sent from a player's computer take an abnormally long time to get to the game server, or server overload, where the server cannot respond to all incoming player commands fast enough, resulting in a noticeable lag time between the command being issued and the results showing on players' screens.

28. Anna Everett, "Serious Play: Playing with Race in Contemporary Gaming Culture." In *Handbook of Computer Game Studies*, ed. by J. Raessens and J. Goldstein (Cambridge, MA: MIT Press, 2005), 311–326.

29. Paul W. Eastwick and Wendi L. Gardner, "Is it a Game? Evidence for Social Influence in the Virtual World," *Social Influence 4* (2009): 18–32.

30. Tanner Higgin, "Blackless Fantasy: The Disappearance of Race in Massively Multiplayer Online Role-Playing Games," *Games and Culture 4* (2009): 3–26.

31. Dimitri Williams, Nicole Martins, Mia Consalvo, and James D. Ivory, "The Virtual Census: Representations of Gender, Race and Age in Video Games," *New Media & Society 11* (2009): 815–834.

32. Garner, *Whiteness*; McWhorter, "Where Do White People Come From?"; Ward, "White Normativity."

33. Garner, *Whiteness*, 37.

34. Steven D. Farough, "The Social Geographies of White Masculinities," *Critical Sociology 30* (2004): 241–264.

35. A. Phoenix, "'I'm White—So What?' The Construction of Whiteness for Young Londoners." In *Off White: Readings on Race, Power, and Society*, ed. by M. Fine, L. Powell, M. Weis, and L. Mun Wong (New York: Routledge, 1996), 187–197.

36. Garner, *Whiteness*.

37. Karyn McKinney, *Being White: Stories of Race and Racism* (New York: Routledge, 2005).

38. Jake Harwood and Karen Anderson, "The Presence and Portrayal of Social Groups on Prime-Time Television," *Communication Reports 15* (2002): 81–97.

39. Kevin DeLuca, "In the Shadow of Whiteness: The Consequences of Constructions of Nature in Environmental Politics." In *Whiteness: The Communication of Social Identity*, ed. by T. Nakayama and J. Martin (London: Sage, 1999), 217–238; Dreama Moon, "White Enculturation and Bourgeois Ideology." In *Whiteness: The Communication of Social Identity*, ed. by T. Nakayama and J. Martin (London: Sage, 1999), 177–197; Judith Levine, "The Heart of Whiteness: Dismantling the Master's House," *Voice Literary Supplement 128* 1(1994): 11–16.

40. Eduardo Bonilla-Silva, *Racism without Racists* (Lanham, MD: Rowman & Littlefield, 2006).

41. Bonilla-Silva, *Racism without Racists*; Cecilia Ridgeway, Elizabeth Heger Boyle, Kathy J. Kuipers, and Dawn T. Robinson, "How Do Status Beliefs Develop? The Role of Resources and Interactional Experience," *American Sociological Review 63* (1998): 331–350.

42. Laurie Taylor, "When Seams Fall Apart: Video Game Space and the Player," *Game Studies 3* (2003), accessed at http://gamestudies.org/0302/taylor/.

43. Lev Manovich, *The Language of the New Media* (Cambridge, MA: MIT Press, 2001); Sherry Turkle, *Life on the Screen* (New York: Touchstone Books, 1997).

44. Specifically, the game had to accept registration from the United States and operate in English. The location of the servers or the game administrators was not relevant, as long as play from the United States was permitted.

45. Many games allowed selection of human or non-human (e.g., Elf, Orc, etc.) avatars, but to make comparisons of virtual skin tone to reality, I disregarded non-human avatars. I believe the implications of similarities in appearance and behavior of non-human avatars to real-world racial groups to be important, but such an examination is beyond the scope of this study.

46. The Von Luschan scale was used in the first half of the 20th century in part to establish racial classifications based upon skin color. See Felix von Luschan, *Beiträge zur Völkerkunde der Deutschen Schutzgebieten* (Berlin: Deutsche Buchgemeinschaft, 1897) and Felix von Luschan, *Voelker, Rassen, Sprachen: Anthropologische Betrachtungen* (Berlin: Deutsche Buchgemeinshaft, 1927). It was unreliable as it depended on the subjective comparison of skin color to a glass color palette by a researcher. Measurements of skin color today typically use reflectance spectrophotometry, but such a method is not suitable for examining computer graphics.

47. See Richard Weller, John Hunter, John Savin, and Mark Dahl, *Clinical Dermatology*, 4th ed. (Malden, MA: Blackwell, 2008).

48. The modification calculates the six categories as follows: I (1–5), II (6–10), III (11–15), IV (16–20), V (21–28), VI (29–36). This simpler categorization is still problematic, as category V has a wide range in terms of darkness of pigment.

49. Erasmus Zimitri, "Hair Politics." In *Senses of Culture,* ed. by S. Nuttall and C. Michael (Cape Town, South Africa: Oxford University Press Southern Africa, 2000), 380–392.

50. Irene Blair, Charles M. Judd, and Jennifer L. Fallman, "The Automaticity of Race and Afrocentric Facial Features in Social Judgments," *Journal of Personality and Social Psychology 87*, 6(2004): 763–778; Irene Blair, Charles, M. Judd, Melody S. Sadler, and Christopher Jenkins, "The Role of Afrocentric Features in Person Perception: Judging by Features and Categories," *Journal of Personality and Social Psychology 83*, 1(2002): 2–25; Robert W. Livingston and Marilynn B. Brewer, "What Are We Really Priming? Cue-Based Versus Category-Based Processing of Facial Stimuli," *Journal of Personality and Social Psychology 82*, 1(2002): 5–18.

51. Of course, the phenotypical traits associated with "blackness" have tremendous variation in terms of skin tone, hair texture, and facial features. Given that the vast majority of games in my study presented "white" characters by default, I judged the capability of the game to create a non-white character by trying to create a "black" character who was phenotypically as far away from "white" as possible. This included dark skin color, hairstyles associated with very coarse hair, and non-Anglo facial features.

52. It should be noted that, in many cases, the art style of the games made it difficult or impossible to distinguish between white and Asian faces. There were seven games that had faces that could be interpreted as Asian, but the determination was ambiguous enough that I do not include them in the main analysis.

53. This is based on number of subscriptions for these games, as reported by the website MMOGData as of August 2009 (see mmogdata.voig.net).

54. Coltrane and Messineo, "The Perpetuation of Subtle Prejudice."

55. Howard Winant, *The World Is a Ghetto: Race and Democracy since World War II* (New York: Basic Books, 2001); Terry Kawashima, "Seeing Faces, Making Races: Challenging Visual Tropes of Racial Difference," *Meridians 3*, 1(2002): 161–190.

56. Williams et al., "The Virtual Census."

57. Dark Legacy Comics, "Whitecraft" (Arad Kedar, 2011), accessed at http://darklegacy-comics.com/190.html.

Chapter Seven

Gendered Pleasures

The Wii, Embodiment and Technological Desire

Adrienne L. Massanari

Women have played videogames since their entry into our homes in the late 1970s and early 1980s, yet the majority of games and game consoles have been designed for young male audiences. Part of this is certainly due to the lack of women working within the gaming industry and the belief that females did not constitute a large enough market to warrant attention from large video game developers and publishers. However, the popularity of games such as *Myst* and *Tomb Raider* in the 1990s[1] and the rise of the casual game movement (as described by Jesper Juul and others)[2] in the late 2000s, attracted new players and demonstrated that men and young boys were not the only ones interested in video games—even if they were still considered the primary demographic for many game companies.

In this chapter, I explore the ways in which Nintendo's Wii console has been designed and marketed specifically for "non-gamers" or those interested in casual gameplay: that is, women and older adults. I explain the ways in which the Wii has been marketed and discursively positioned to configure the domestic sphere in particular ways. In addition, I examine the ways in which Nintendo's unique motion controller (the Wii remote) encourages players to use the Wii remote as a prosthetic device that engages the body in play in new ways. I also explore Nintendo's fitness game, *Wii Fit*, and describe how both the game's marketing and the playing experience it offers, discursively reinscribes certain stereotypical gender politics and "disciplines" our bodies in new ways. I argue that *Wii Fit* proscribes a particular view of pleasure, one that focuses on a potentially problematic perspective of what it means to be "fit." To start this exploration, I trace the ways in which scholars and popular culture have considered the body (particularly female bodies) as it relates to technology and games.

TECHNOLOGICAL DISCOURSES OF EMBODIMENT

Feminist scholars have long placed the body at the center of their inquiry, because "there is a tension between women's lived bodily experiences and the cultural meanings inscribed on the female body that always mediate those experiences."[3] Scholars in other fields have also wrestled with the "body problem." For example, those working within cyberculture and new media studies have struggled to define and describe what happens to the body when we go online. Early visions often lapsed into utopian fantasies wherein the physical body would be left behind for the seemingly pure world of cyber-space, where—like the famous MCI "Anthem" commercial suggests—we would communicate "mind to mind," free from "infirmity" and whatever had previously divided us.[4] Such was the fantasy expressed in both William Gibson's *Neuromancer*,[5] in which the term "cyberspace" was coined, and Neal Stephenson's *Snow Crash*,[6] which extolled the social freedom afforded by the "metaverse." Early academic discourse regarding these technologies also reinscribed the belief that the body, if not left behind entirely, could be expressed in a multitude of ways; and that we could transcend the gender, racial/ethnic, and social divides that had separated us in the past.[7] For example, Donna Haraway's idea of the cyborg—a human-machine hybrid—emphasized the libratory potential of technology to dissolve or challenge gender binaries.[8] At the same time, the appearance of cyborgs in popular culture suggested another discourse in which the body (usually male) was an object of physical strength and emotional fortitude. Science-fiction films from the 1980s such as the *Terminator* series,[9] *Robocop*,[10] and *Total Recall*[11] directed the viewers' gaze towards the hypermasculinity of these half-men/half-ma-chines. Cyborgs in these films were to be feared and respected.

Later, stories of technologically augmented bodies became more nuanced. In films like *The Matrix*[12] and *eXistenZ*,[13] the body/"real world" becomes a yoke to shrug off in favor of the virtual. In addition, the act of entering the virtual space is framed as both uncomfortable and yet pleasurable. In *eXis-tenZ*, for example, Ted Pikul (played by Jude Law) mentions his concern about having a "bio-port" placed in his lower back—that he is unsure about allowing himself to be penetrated by the organic, umbilical-like cord needed to play the game. In *The Matrix*, Neo (played by Keanu Reeves) must con-nect to the virtual world through a port in the back of his neck. His first experience with "jacking in" elicits a sigh of surprise (and presumably pleasure) as he experiences freedom from his corporeal body. The films cannot help but imply a kind of erotic pleasure in leaving the physical realm for the virtual. As Claudia Springer notes in her interrogation of discourses about the cyborg, "computer technologies . . . occupy a contradictory discursive position where they represent both escape from the physical body and fulfillment of erotic desire."[14]

Around this time, new media scholars suggested the relationship between the "real world" and the virtual was far more nuanced than these early utopian discourses suggest. Race, gender, class, sexuality, and (dis)ability did not simply disappear when interacting online; rather, these realities of our everyday world shaped (and were shaped by) our entry into cyberspace.[15] Indeed, the very interface through which we engaged virtual/computerized environments often made us subjects to a complex network of power relations.[16] In cyberspace, identities could be more fluid than in non-virtual spaces; however, the social-cultural structures that shaped our experiences in the everyday world did not cease to shape them online. Instead, our interactions were shaped by the choices designers and producers make when creating a technological artifact—and these choices were further influenced by professional design practices, organizational politics, and economics.[17]

Video games and synthetic worlds[18] further complicate this complex relationship. By introducing another factor—the avatar, or, "an interactive, social representation of a user"[19] —questions of identity and representation are raised. As Adriano D'Aloia argues, the avatar "is an extension of the player's own body, and entertains a prosthetic relationship with it; it incorporates the player and disciplines his/her body. It is the embodied manifestation of the player's engagement with the game-world; it is, at the same time, a reflection of ourselves and an envoy of ours in the parallel world."[20] Through the avatar, we experience these spaces through a representation of our selves— one often not of our own choosing. T. L. Taylor suggests that avatars are important sites, "through which users not only know others and the world around them, but themselves."[21] These often graphical representations force us to consider complex questions regarding what it means to be present and autonomous in the game space.

WOMEN, TECHNOLOGY, AND GAMING

Gender is a complex social and cultural construct and fully interrogating the interplay between gender and technology is beyond the scope of this chapter. However, it is worth briefly exploring what theoretical grounds guide my work in this area. My thoughts about the relationship between gender and technology are shaped by Teresa de Lauretis' argument that, "the sex-gender system . . . is both a sociocultural construct and a semiotic apparatus, a system of representation which assigns meaning (identity, value, prestige, location in kinship, status in the social hierarchy, etc.) to individuals within the society. . . . The construction of gender is both the product and the process of its representation." [22] Judith Butler suggests a complex interplay between the body and biological sex, such that,

> "Sex" is an idealized construct which is forcibly materialized through time. . . . Once "sex" itself is understood in its normativity, the materiality of the body will not be thinkable apart from the materialization of the regulatory norm. "Sex" is, thus, not simply what one has, or a static description of what one is: it will be one of the norms by which the "one" becomes viable at all, that which qualifies a body for life within the domain of cultural ineligibility. [23]

As with gender, technology is a socially constructed object that both shapes and fits into existing social practices. Early scholarship grounded in this perspective—the social-construction of technology, or SCOT [24] approach—explored the relationship between women and technology often focused on the role of domestic technologies in women's lives, and how it reinforced clear delineations between the "domestic" (feminine) and the "public" (masculine) spheres. Later investigations of gender and technology using this theoretical framework interrogated specific technologies and the social practices surrounding them. [25]

As with many other new technologies, early popular discourse about video games suggested that games and gameplay were a masculine pastime, one of little interest to women or girls. [26] In the mid-to-late 1990s, the "girl games" movement questioned this assumption, and argued that it was imperative to encourage gameplay among girls as such formative experiences could pique their interest in technology, science, and math generally, and ensure they would gain important technological literacy skills. [27] As Henry Jenkins and Justine Cassell note in a follow-up volume to their influential *From Barbie to Mortal Kombat*, there were several competing goals that drove the girl games movement: economic (as young male players were already well-served by the current crop of games and consoles); political (as females were lagging behind males in science and technology fields, which was attributed to the fact that fewer young girls engaged with computers the way that their male counterparts did); technological (as multimedia CD-ROM drives became standard on PCs, which meant that families did not need to purchase a separate console for their children to play games); entrepreneurial (as more women were starting female-focused gaming companies); and aesthetic (as their was an increasing desire for richer stories and different interfaces that might appeal to a broader spectrum of gamers). [28] Proponents of the movement, such as Purple Moon's founder Brenda Laurel, suggested that girls were interested in different types of stories and challenges than boys were.[29] In particular, designers were encouraged to incorporate more social and less violent activities in their games.[30] While the suggestion that girls and boys differed in their approach to gameplay may have been accurate, suggesting that "pink games" should be designed around girl-centric content (like shopping, socializing, and domesticity), perpetuated an unfortunate binary where girls who preferred *Halo* to *Barbie Fashion Designer* were perceived as exceptional cases or outside the norm.[31]

While females' interest in games is finally being acknowledged within the mainstream press, the discourse around gaming still relies on outmoded binaries to describe the pleasures that men and women take when playing video games. For the most part, the media and popular culture still employ the term "gamer" (or "hardcore gamer") to refer to the core demographic for which many games are designed: young males (under 30), who have a substantial amount of time to master a game's content, money to spend on game consoles and new games, and are (presumably) interested in militaristic, competitive games in which females are either passive objects of sexual desire or absent from the game's narrative altogether.[32] This is despite evidence to the contrary that suggests the average "gamer" is as likely to be female as male, and that numerous players are over the age of 30.[33]

In their research into the ways in which games might be designed to better facilitate girls' wants/desires, Cornelia Brunner, Dorothy Bennett, and Margaret Honey suggest that men and women differ in their technological fantasies.[34] They argue that women view technologies as something to fit into their already-existing surroundings and relationships, in opposition to men who typically view technology as a tool for efficiency and power.[35] While the reductionist nature of these desires may be problematic (as noted by T. L. Taylor),[36] the notion of "technological desire" is a fruitful way to examine the relationship between women and gameplay.

One of the criticisms leveled at much of the early research into female gameplaying is that it conflated the needs and desires of girls with those of women. As Pam Royse and her coauthors note, rarely do studies problematize this approach.[37] More recent research has focused on adult female players, and suggested that much of the notion that women are most interested in aspects of "identity play" while gaming reifies and reinforces the idea that women are somehow viewed as "intruders" into presumably male-dominated gaming cultures.[38] As Elisabeth Hayes suggests, while gender should not be ignored when thinking about games and the practices of game players, simply categorizing games based on their likelihood to appeal to males or females is problematic. "Female" gaming practices are highly individual, and are influenced by prior experiences with games, individual identity and self-formation (especially when it comes to more "gendered" gaming experiences or genres like fighting or first-person shooters), and evolve as players become more practiced.[39]

Unfortunately, game designers often rely on out-dated or over-simplified dichotomies when thinking about women's play and pleasure when gaming. For example, Jesse Schell's *The Art of Game Design*[40] wisely suggests thinking about the audience who will be playing the game while designing it. This potentially represents some progress within the gaming industry, which, like other fields populated by technically adept individuals (such as computer programming), tends to look inward as a guide for what others might like to play. Unfortunately, Schell offers only broad stereotypes when describing what women find attractive in games. He suggests they prefer "experiences that explore the richness of human emotion," "entertainment that connects meaningfully to the real world," "nurturing," "dialog and verbal puzzles," and "learning by example."[41] Further, he suggests that men, "enjoy mastering things," "competing against others," "destroying things," "spatial puzzles," and "trial and error."[42] Such simplified taxonomies of difference may serve to further disenfranchise both women and men from many mainstream video games, especially if these are the only exposure the would-be game designer has to the complexities of player desire. In addition, if game developers continue to simply reproduce the kinds of games they themselves would like to play, and the industry continues to be dominated by mostly white males in their late 20s and early 30s, most games will continue to reinscribe male, heteronormative, white, and middle class ideologies.[43]

YOU AND MII = WII

Nintendo released the Wii console in 2006 as a follow-up to the relatively disappointing sales of their GameCube. By March 2009, Nintendo had sold almost 26 million Wiis worldwide.[44] One of the differences between the Wii and other next-generation consoles is that the Wii lacks many of the high-end graphics capabilities of the Xbox 360 or PS3. While the same game might be released on each console, video game enthusiasts inevitably decry the Wii's graphics as a "watered down port of the original."[45] In its original review of the console, gaming blog "Ars Technica" argued that the "graphical prowess" of the Xbox 360 and the PS3 were their primary draws, suggesting that the Wii's main attraction was its "innovative control scheme."[46] Thus, the Wii was (and is) positioned as the "family console"—one that, like Nintendo's earlier handheld (DS), might draw new players who do not view themselves as "gamers."

The most unique aspect of the Wii is its use of a wireless motion-sensing game controller. The Wii remote (or "Wiimote") uses a mimetic interface, where avatar movements correspond to the movements that the player makes using the device. It encourages intuitive physical action in lieu of complex button sequences to move the player's character. For example, to bowl in Wii Sports[47] a player holds the Wiimote as if she is holding a bowling ball and releases the button on the back of the controller as she swings her arm forward. As Mihaly Csikszentmihalyi argues, physical movement (like that enabled by the Wiimote) often encourages individuals to lose their sense of self-consciousness and enter what he terms a "flow state," heightening their pleasure and enjoyment.[48] Additionally, mimetic interfaces like these encourage interactions in the player space—that is, we focus on the interplay between the physical gaming environment and the screen.[49] The Wiimote takes on various roles depending on the game, becoming a tennis racket in Wii Sports, a guitar in *Guitar Hero*, or a steering wheel in *MarioKart*. While marketing campaigns for the Wii show players standing up, swinging their arms like they are serving a real ball on a tennis court, players quickly figure out that flicking one's wrist in the same downwards motion, while sitting on the couch, can have the same effect.[50] Therefore, the Wii offers an imperfect simulation of what it means to play tennis (or play guitar or race cars). Jesper Juul argues that it is precisely this imprecision that allows novice players to feel competent and engaged, and encourages participation from individuals who are unlikely to have owned or played other next-generation consoles. At the same time, the imperfect simulation provided by the Wiimote may frustrate expert tennis (or guitar) players, as it does not allow for the kind of fine-grained control to which they are accustomed.[51] However, the Wiimote's easy-to-use interface encourages most individuals to feel competent while playing—and likely attracts players who may feel overwhelmed by the numerous button combinations required to play most Xbox 360 and Playstation 2 and 3 games.

This duality—the Wiimote's ability to allow novices and non-traditional game players to feel competent when they play, and the "imperfect" simulation it offers—may also explain the conflicted relationship "hardcore" gamers have with the Wii. Initial press coverage of the Wii by the gaming press was positive.[52] However, there were a number of individuals who felt the Wii was not worthy of the "next-generation" title. For example, Chris Hecker, a developer working at *Maxis*, ranted at the 2007 Game Developers' Conference, arguing that "the Wii is nothing more than two GameCubes stuck together with duct tape."[53] Gaming magazine *Edge* noticed that game review scores for those released for the Wii console tend to be lower than the Xbox360 or PS3, suggesting that it might have something to do with the lower quality graphics and hardware which might lower the barrier to entry and encourage many more sub-par games to be released for the console. At the same time, they also noted that the innovative interface the console allows might be more difficult for game designers to work with. As the author notes, the Wii is "innovative but low-tech; it's accessible for gamers, but difficult to nail from a game design perspective; it has a large install base, but one that has proven tough for third-parties to crack."[54]

Besides the innovative Wiimote controller, other aspects of the Wii were designed to appeal to a larger audience. Before playing any games on the Wii, players are encouraged to create an avatar, called a "Mii," to represent themselves in various games.[55] Miis are broad, non-realistic caricatures. Mii heads are big, with eyes that borrow heavily from the Japanese anime traditions, whereas their bodies are small and relatively shapeless. Instead of hands, Miis have round knobs, and their clothing is limited to a simple shirt and pants—the only exception is that in some games, special outfits can be unlocked.[56] Miis can be male or female (gender is specified on the first screen of the Mii creation process), but this does not preclude the player from creating a relatively androgynous-looking Mii, as hairstyles, facial shapes, and other attributes are not limited based on the gender choice. Optionally, Miis can travel across connected Wii consoles, which means that individual players can access and play games using Miis created by their friends or family. In addition, Mii characters are often present in many different Wii games. For example, in Wii Sports, Miis appear as spectators, cheering on the player's Mii. In Wii Fit, Miis serve as backdrop characters—alternatively tossing hula-hoops to the player, jogging beside them, or kicking soccer balls for the player to hit. In this way, the Wii encourages some sense of belonging to a social, collective sphere even if the player is physically alone.

THE WII AND DOMESTIC PLEASURES

The Wii's small and light design is a radical departure from the dark and hulking forms of the Xbox and Playstation. In an interview on Nintendo's corporate site, designer Kenichiro Ashida suggests that the Wii was envisioned as being able to seamlessly integrate into any household's living room. He notes,

> We came up with "A Design for Everyone," a concept created in order to allow as many people to use Wii as possible. Making Wii into a device that everyone likes is more important to us than having a fiercely individualistic design. Indeed, we wanted to make Wii into something that would be treated more like a piece of interior design, rather than a toy or a piece of AV equipment.[57]

In their attempt to create a console that is for "everyone," Nintendo's team implicitly references a familiar criticism leveled at the gaming industry—that consoles (and games) are most often created and marketed to young males who enjoy photorealistic graphics and exciting (read: violent) gameplay. In some ways, the use of the term "everyone" is really a coded way to suggest that the Wii will fix the industry's supposed inability to attract female and older gamers.[58] Additionally, the notion that the console is an addition to a living room's interior design plays on stereotypical notions of femininity and female control of the domestic sphere. However, such a statement suggests somehow that women are not already playing games—a point that has been debunked by a number of different scholars.[59] Bonnie Ruberg, writing for the popular gaming blog "Joystiq," argues that while Nintendo's interest in women is admirable, their marketing campaigns often go too far:

> On the one hand, it's refreshing to see a major player like Nintendo thinking about women—not just in terms of one game, but a whole console, and with it a slew of "non-girly" titles. It's also encouraging to see female players linked with innovation, something the video game industry as a whole needs desperately. Women have finally made it onto the larger marketing map. . . . At the same time, some female gamers are understandably bothered by claims [that the Wii is attracting new players—females and older adults] like the ones [President of Nintendo America Reggie Fils-Aimé] made. . . . First off, women players already do exist; we're right here. It's just that, until we bring in the big bucks, we don't seem to matter. Second, women are people, full grown adults who can make decisions for themselves about what they like or dislike—video games included. Telling them what they'll play, so the argument goes, is insulting to their ability to make choices.[60]

Ruberg continues by noting how Nintendo's advertising campaign, while showing many women playing the console, they are surrounded by other people—family members, partners, children, etc. Thus, it configures what constitutes "appropriate" female gameplay as a social activity.

The Wii penetrates the domestic sphere in unique ways. Unlike the Xbox 360 or Playstation 2 or 3 which might be relegated to the teen boy's bedroom, the Wii is imagined as a part of the family room space. As Bernadette Flynn argues, game consoles have been "domesticated"—entering into the living room space in the ways that older technologies, such as the radio and television have in eras past.[61] Still, she notes that there is an inherent contradiction between the ways in which Sony and Microsoft discuss their consoles' centrality in the living room and the ways in which these consoles are advertised. She writes,

> There is little attempt by video console manufacturers and distributors to present the video-game console as a domesticated object. Instead, commercial advertisements emphasize the game console as a futuristic dream machine in opposition to the place of its location—suburbia and, specifically, the living room. . . . Advertisements for video games draw on these modernist notions of the home where the monotony of media consumption is transformed into the

fantasy of an electronic portal to a virtual exterior. The digital hearth with its flow of data transforming play into work and work into play evokes images of a fun palace away from the mess and routine of digital life in the home. [62]

Other scholars note that the sociable play encouraged by consoles, and the Wii in particular, encourages its positioning within the home's shared living spaces. [63] It is also important to note, that part of what encourages the console's position in the home has much to do with the audio-visual equipment that these devices require to play, which are often located in the living room or other shared spaces. [64]

As the game console enters the domestic sphere and becomes a primary part of many central living spaces, a corresponding set of social practices arises. These practices often suggest a conflicted relationship between pleasure, play, and sociability. In a study in which she examines how men and women "rationalize" gaming to themselves and others, Helen Thornham argues that there is a certain stigma of taking pleasure in the game for the game's sake. [65] Instead, games are "placed into defining parameters of the social (only in a social situation can you enjoy and take pleasure from gaming), where the pleasure of gaming is less about the games themselves and more to do with the presence of friends in a social environment." [66] Further, she argues that social gaming is often equated with "normal" gaming, whereas solo gaming was perceived as a "geeky" pursuit. [67] It is perhaps not surprising, then, that Thornham discovered that women are far more likely to engage in socializing behaviors during gameplay. [68] For example, she observed that female game players may actually downplay their expertise when playing games with males, which may have more to do with their socialization as females than their actual experience playing games. Recent observational research with Wii players suggests that at least some women often only play console games with others. Gaming, in their minds, was purely a social activity to be enjoyed when others were physically present. [69] Other studies have found that women very rarely play games with same-sex friends; rather, they are far more likely to play with opposite-sex friends or partners. [70] This suggests that females have a complex and often contradictory relationship with the pleasures of gameplay, especially as it is inseparable from the network of social relationships and gender politics that define and are often reinforced through the domestic space of the living room.

Such findings are supported by many decades of work within cultural studies of the media, which attempt to understand and link texts and audiences and how they are both influenced by economic power and ideology. While the concept of "pleasure" has been invoked in many of these studies, scholars suggest that it remains under-theorized. [71] Certainly, part of the pleasure of engaging with the media is our ability to make meaning, as John Fiske has argued. [72] But just how this kind of "meaning making" shapes and is shaped by the social world is far more difficult to unpack. Especially complex is the notion of "women's pleasures" and their relationship to media texts, as Barbara O'Connor and Elisabeth Klaus argue. They note two overarching trends within most media/cultural studies scholarship of women's media pleasures: such pleasures are positioned as overwhelmingly positive and potentially resistive (especially given the frequency by which media directed towards women are derided in popular culture), and most often, studies of these pleasures do not fully unpack the relationship between pleasure, politics, and ideology.[73] Contemporary studies of digital media suggest our pleasure using these new forms is highly individual, but that they involve some enjoyment of the key elements of new media: interactivity, intertextuality, and control.[74]

CASE STUDY: WII FIT AND WII FIT PLUS

In 2008, Nintendo released Wii Fit, an exercise game bundled with a "balance board" that used the Wii's motion-sensing technology to determine the player's position.[75] The white balance board is about 20 inches long, 12 inches wide, and around 2 inches thick and wireless connects via Bluetooth to the Wii. The balance board can calculate the mass of a person standing on it and uses four sensors to determine to what direction that person is leaning.[76] Much like the Wii itself, the balance board was conceived as a platform that would be useable with many different games. As of this writing, a number of games have been developed specifically for the balance board (such as *EA Sports Active* and *Jillian Michaels' Fitness Ultimatum*), or can be played using the balance-board instead of the Wiimote (*Mario* and *Sonic at the Olympic Winter Games*, *We Ski*, and *Punch-Out*, among others).[77] The bundled game, *Wii Fit*, includes yoga poses, strength and aerobic exercises, and balance games. Supervising your progress throughout the game is a male or female personal trainer (your choice), whose encouragements ("Great job!") and admonitions ("You're wobbling . . . ") change based on your performance as the screen displays an image of them completing the exercise/pose with you. Although the game starts with only a few poses and exercises available, time spent playing eventually unlocks additional strength workouts, balance games, and aerobic activities.

Wii Fit Plus was introduced in 2009. Like *Wii Fit*, the game is bundled with a balance board, or can be purchased alone for those who already own the original software. *Wii Fit Plus* updates the original *Wii Fit* game, including new multiplayer games, fitness tracking for pets and young children, and a customized workout routine.[78] Before beginning either version (*Wii Fit* or *Wii Fit Plus*), players are asked their age and height, and are encouraged to complete a test to evaluate their fitness level. While standing on the balance board, the player is weighed and tested on their balance and posture. *Wii Fit* then calculates the player's body-mass index (BMI) and unveils, with much fanfare, her "Wii Fit Age." In addition, players can set a goal weight loss and a date by which they would like to achieve their goal. This information, along with the player's "Wii Fit Age" is tracked over multiple sessions.

WII FIT AS A CASUAL GAME

Casual games are "video games developed for the mass consumer, even those who would not normally regard themselves as a 'gamer.'"[79] Traditionally, casual games have attracted many players who would not otherwise play games—that is, they are older and more likely female than other kinds of video game players. Casual games are often contrasted with so-called hardcore games, require less time to play and typically a less-intensive learning curve.

Jesper Juul argues that casual games typically share five elements of design: a positive fiction, a highly usable interface requiring little instruction to begin playing, gameplay that allows for or assumes interruptions, game progression that challenges players but does not force them to replay levels, and positive feedback, which he terms "juiciness."[80] Using Juul's taxonomy, it is clear that *Wii Fit* is intended to be viewed as a casual game, but there are elements that suggest it defies easy classification as such. In terms of the game's story or fiction, it takes place in the same cartoonish world of Wii Sports. Many of the games take place in brightly lit, social environments—for example, the aerobic running game allows players to run their Mii in a park, surrounded by other Miis representing family members and friends. However, the overall purpose of the game, getting fit, may not be viewed as entirely fun or friendly, especially as the "Wii Fit Age" and use of BMI and balance as fitness indicators (mentioned above) may inadvertently disenfranchise certain players. *Wii Fit* is designed for interruptibility. The game's activities are short—ranging from a minute to ten, suggesting that players can easily stop and start gameplay as necessary. However, the game is also predicated upon regular, consistent play. Players receive negative feedback from the game if they do not play for a number of days, or play only for short periods. This is considerably different from other casual games, where the game does not refer to the last gameplay session, other than to preserve high scores over time.

Unlike some other casual games, players have much more control over the individual challenges—as they are able to choose the difficulty and number of repetitions of certain exercises in the game. At the same time, players are encouraged to repeat the games and exercises both for intrinsic rewards (for example, to move up in the "high score" list for each game and to unlock new activities), and extrinsic ones, as presumably their fitness levels improve if they consistently complete the exercises. In this way, *Wii Fit* differs from other casual games.

Positive feedback, or "juiciness," abounds in *Wii Fit*. The player's personal trainer offers positive reinforcement during the strength exercises and yoga poses. In the balance and aerobic mini-games, players are ranked at the end of their session and musical fanfare accompanies this announcement. However, unlike many games that demonstrate juiciness, *Wii Fit* is not always positive about the player's performance. For example, performing badly on a particular exercise leaves one's Mii hanging her head in shame.

HOW *WII FIT* DISCIPLINES THE BODY

Both *Wii Fit* and *Wii Fit Plus* fit into the "exergaming" (exercise gaming) genre, most famously represented by Konami's *Dance Dance Revolution* (DDR) series. While not designed as health games per se, these kinds of games are seen as a possible solution to the increasing rates of obesity, particularly among young people.[81] Ian Bogost traces the roots of the exergaming genre to the original cabinet video games of the 1980s such as *Ms. Pac-Man* and *Donkey Kong*, which required standing up to play, Atari 2600's "Foot Craz" controller released in 1987 that had five buttons that players could use in lieu of the joystick/button controller in certain games, and Ninetendo's NES "Power Pad," released in 1988, which could be used with games like *World Class Track Meet* in which players completed track-and-field sporting events.[82] He suggests that most exergaming games use several different rhetorical appeals and tactics to encourage continued play and retain the player's interest in the physical activity enabled by these games. Typically, these appeals are made through different play styles: sprinting (through games that require players to physically run in place—mimicking track racing); agility (in which players quickly change between running in place and jumping to mimic hurdle jumping, for example); reflex (activities in which players have to mimic certain movements); training (activities that mimic physical workout training and the player's progress is tracked over time); and impulsion (represented by games such as DDR, where exercise is "emergent" and a secondary goal of play).[83] In addition to these rhetorical appeals, researchers have argued that design choices also make a difference as to the effectiveness of these kinds of physical fitness technologies. These include: ensuring that individuals receive credit for the activities they do, as well as giving them a clear sense of how they are performing; encouraging social influence through pressure and support; and enabling customization or consideration for each individual's particular lifestyle.[84]

Especially during its initial release, many health professionals praised *Wii Fit* for its potential benefits, and that its fun approach might appeal to individuals who would be reluctant to exercise on their own.[85] However, there are several potential issues with the ways the game assesses fitness. First, the "Wii Fit Age" is based predominately on the results of the initial balance test (according to the instruction manual packaged with the game). Since part of the balance test requires being familiar with the balance board's unique control scheme, it is likely that first-time players will find themselves being marked as older than they actually are, which can be frustrating and a bit shocking (as the author can attest). Interestingly, balance is not included in either the fitness tests offered by the U.S. Department of Health and Human Services' President's Council on Physical Fitness and Sports[86] or the U.S. Army.[87] Both these tests focus instead on an individual's performance of a number of aerobic activities such as push-ups and running. Second, the game requires you to weigh in to receive a "stamp" (as a mark of completing your fitness activities for that day) regularly. Studies with young women have shown that daily weigh-ins can actually encourage obsessive behavior, and may lead to unhealthy weight-management practices (like taking laxatives or skipping meals).[88]

Another potential issue with *Wii Fit* is its use of the BMI to assess the player's body fat. This scale measures the ratio of a person's weight to their height and the resultant number indicates whether the individual is considered underweight, normal weight, overweight, or obese.[89] However, the BMI has been criticized for incorrectly categorizing individuals who are more athletic into the overweight category, as muscle weighs more than fat.[90] In addition, individuals with BMIs that suggest they are overweight may actually be at a lower risk for heart disease than those who fall into the "normal" category.[91] Some feminists have also suggested that the BMI is an inaccurate measurement of health and perpetuates a negative body image[92] —one blogger has gone so far as to solicit pictures from women with their height, weight, and BMI, which she then posted to Flickr to challenge how the BMI categorizes real women's bodies.[93] *Wii Fit*'s use of BMI as the primary way to assess fitness, therefore, is problematic.

Despite positive initial reactions from those who thought *Wii Fit* might encourage sedentary individuals to become more active, more comprehensive testing of the game suggested that it burns fewer calories than the *Wii Sports* game that comes bundled with consoles sold in the United States.[94] The American Council on Exercise (ACE) suggests that *Wii Fit* yields "underwhelming" caloric expenditure and involves less intense exercise than expected—even if players are engaging in the most aerobically challenging activities the game offered (the running, stepping, and hula-hooping activities).[95]

Michel Foucault's work on how power is enacted through discourse is particularly salient when examining how the body is experienced/configured when playing *Wii Fit*. Foucault argues that scientific power (enacted through technologies) creates "docile bodies" which are far easier for social institutions and individuals holding power to control.[96] Certain choices made by the designers of *Wii Fit* necessarily discipline the body in particular ways. Consider, for example, the *Wii Fit* fitness test players are encouraged to participate in before starting to work out. One of the more stunning ways that the game "disciplines" players' bodies occurs when they receive their BMI score. If the player's BMI falls into the upper-level categories (overweight or obese), their Mii is distorted and morphed as the game tells the player that they need to lose a certain number of pounds before their Mii will return to its original shape. Thus, the player's self-representation/avatar is no longer under her own control. While these changes do not persist across games on the Wii console, the experience of having one's Mii "fattened up" without permission is disconcerting.

Other choices made by the game designers discipline the body through surveillance. For example, if multiple people have registered their Miis with the game, and one logs in more regularly than the other, the game says, "You know, it's been a long time since I've seen [name]." If one has not completed any activities for a while, the game greets you with this fact, noting, "Nice to see you again! It's been [x] days since I saw you last!" While such reminders might help one stay motivated, it feels an awful lot like one big passive-aggressive guilt trip.

In addition, the game creates a sort of enforced sociability. Even when playing "alone" other Miis join in, tossing hula-hoops or soccer balls, or doing laps beside you while you run. They turn their heads when they pass you, and cheer and wave when you run by. At no time is a player "alone"—even when performing the yoga poses and completing the strength exercises, one's personal trainer appears on the screen to encourage or chastise.

If a player does not perform activities correctly, *Wii Fit* is quick to punish, and it does so most publicly. Run too slow and you are encouraged to speed up. Run too fast, and your Mii humiliatingly trips and falls. Failing at a balance game causes your Mii to hang her head in shame. The top ten scores for all activities are shown at the end of each game and a triumphant fanfare or a much less enthusiastic few notes accompany this feedback. Again, these choices may motivate some individuals to exercise, but they also clearly discipline the player's body in particular ways.

ENVISIONING FEMININITY THROUGH ADVERTISING FOR THE *WII FIT PLUS*

Advertising often plays a key role in both educating the public about how to use new technologies and discursively constructing appropriate use of these technologies.[97] In the contemporary media environment, however, advertising is less about selling a product and more about selling a lifestyle, and so, "ads do not focus on the argumentation and evidence of why the product is good but rather on creating situations in which a consumer is satisfied and happy when using the product."[98] As Althusser argues, advertising, like other ideological state apparatuses, interpellates viewers as subjects.[99] While audiences are not passive and resistive readings of media messages are possible,[100] advertising often highlights intended uses and audiences.[101] And, since advertising is often the first time audiences become aware of a product, it will likely influence on some level what they think the product is about and for whom it is intended.

Researchers have noted the gendered nature of most video gaming advertising, as it tends to focus on male fantasies of in-game violence and often stereotypical representations of women and men.[102] In contrast, Wii advertisements typically feature people, rather than focusing on the gameplay itself. As one scholar notes, "With the Wii, selling the console and its games, indeed selling gaming, has shifted from the promise of virtual world experiences on the screen to the promise of the experience of players in the living room. The object of consumption is no longer just the spectacle of the game on a screen but rather players' corporeal engagement and kinaesthetic involvement in that spectacle."[103] Therefore, examining marketing material for the *Wii Fit* and exploring the ways in which it represents pleasure and play is vital.

Two recent advertisements created by Goodby, Silverstein & Partners premiered in the United States as part of Nintendo's 2009 holiday ad campaign promoting *Wii Fit Plus*.[104] Both commercials feature a side scrolling, single take of a "day in the life" of a white, presumably middle-class (but actually quite affluent) mother. The first, "Working Mom," starts with an alarm ringing, a woman getting out of bed as her husband rolls over. We then see her pick out an outfit from her closet, throw something in the hamper in her bathroom as she brushes her teeth, and move to the kitchen (dressed now in workout clothes), where she catches some toast that pops up, and tosses it to her daughter while saying, "Hey, sweetie!" She then blows a goodbye kiss to her husband (dressed in business attire) and moves to the living room, where the commercial slows down (and birds chirp happily in the background) as she performs a yoga pose in front of her television using *Wii Fit Plus*. We then hear an elevator ding as the commercial speeds up again, and now our working mother enters a chaotic office scene dressed in business wear, answering the phone with a stern, "Approved," and clicking the next slide of a presentation in front of a group of colleagues. The scene then changes briefly to a grocery store, where the working mom says a cheery, "Thank you" to the checkout person and grabs a bag of groceries, finally places them down on a chair in her living room, and sits down next to her husband as her young daughter plays *Wii Fit Plus*. The female announcer says, "Custom workouts and fun new games. Fit some fit in, with *Wii Fit Plus*" as the words, "*Wii Fit Plus*" overlay this bucolic scene. The commercial ends with a plug for the Wii console's new price of $199. [105]

The second commercial, "New Mom," is much like the first, featuring the same jaunty music and rapid pacing. Instead of an alarm, however, a baby's cries open this commercial, as our new mom jumps out of bed, crosses to the baby's room, turns on the light, and picks her up with a kiss. We then hear a telephone ring and she enters the light-filled kitchen to answer it, still holding her child in her arms. As she crosses the room, we see a plumber working on her sink, and she hands the baby over to her husband, who kisses her, while the woman leaves the scene with a "Bye" and a wave. She then appears clad in workout gear in her living room, as the commercial slows down, boxing to *Wii Fit Plus*. Then, a series of playing children enter the frame, and we see her (in casual clothing) greeting her husband with a kiss and taking a stroller from his hands, which she then pushes into a dry cleaner and receives a shirt with a "Thanks." The scene then changes—the stroller has now become a vacuum cleaner, and we see our new mom finally collapsing on the couch as her husband plays *Wii Fit Plus*. Again, the commercial ends with the same tagline, "fit some fit in," and a brief promotion of the Wii console's new low price. [106]

These commercials suggest a number of things. They both leverage a highly stereotypical view of the modern woman. First and foremost, she is a mother—one who is overworked (either balancing family/home and work or serving as primary caretaker of a new child). We see her performing a majority of the domestic chores—be they grocery shopping, cleaning, or child rearing. In the "Working Mom" sequence, we see her in a position of power in the workplace (as evidenced by her stern, "Approved" into the phone), but she is still the one shopping and tending to her daughter's breakfast before she works out. In "New Mom," we see a slightly more equal sharing of the child care routine—the husband takes her daughter out for a walk while she uses *Wii Fit Plus*, but she is then expected to pick up the dry cleaning and vacuum while he presumably plays with the Wii or works. Thus, both commercials posit an unequal sharing of domestic and child caring responsibilities. Second, in both sequences, there is a chronic sense of "busyness"—and playing *Wii Fit Plus* is presented as a remedy and a respite from the woman's everyday life. It is the only moment she is alone and presumably relaxed—as evidenced by the fact that the camera's movement stops tracking the woman (albeit briefly). The room in which she uses *Wii Fit Plus* is brightly lit, simply but comfortably furnished, and seems as a sanctuary away from the rest of the home's chaos. Interestingly, in both commercials, this room is different then the living room we see in the final family *Wii Fit Plus* sequence—suggesting this is a "room of her own." Presumably, our harried mothers need to, "fit some fit in" otherwise they will not work out on their own or take time for themselves. Third, both commercials reflect on the sociable nature of play. In both, the mother's involvement with *Wii Fit Plus* occurs only when she is alone, as her daughter in "Working Mom" and husband in "New Mom," respectively, are the ones playing in the final sequence when the entire family is together. In the solo sequences, the mother is doing yoga or boxing—truly "working out." In the family sequences, the daughter and husband play *Wii Fit Plus'* balance minigames. Thus, these commercials reinscribe a stereotypical notion of pleasure and play. The mother is not "playing" when she is using the game alone—she's working out and taking time out from her busy life. She is actively working to achieve her fitness goals. At the end of both commercials, the mother moves from active participant to passive spectator, as she watches her husband (daughter) play the game. This suggests that pleasure in play (as opposed to working out) occurs in the company of others. Interestingly, also, is that this reinscribes the distinction Thornham makes between what constitutes "social" versus "hardcore" or "geeky" gameplay.[107] Additionally, the commercials suggest that it is the wife/mother who needs to be using *Wii Fit Plus* to get fit—not the husband or other household members.

What is particularly interesting about these two commercials is how limited a view they offer on the possibilities of *Wii Fit Plus*. In other, non-U.S. markets, a broader range of activities and individuals are shown using the game. For example, a spot aired in Spain shows a woman playing the hula-hooping minigame and completing one of the strength exercises while a man (presumably, her partner) looks on.[108] This same couple is featured in other commercials that aired—in some, the woman looks on as her partner exercises (and they laugh as he completes some of the activities;[109] in others, she tells him about the game as she completes one of the yoga poses.[110] A Korean commercial features a rotating group of individuals boxing—first we see a woman playing while her two male friends look on, then two young men, then a mother plays while her son watches, then two young women play, then finally a man plays while his wife and daughter watch.[111] Interestingly, all of the spectators in this spot also perform the activity, even though only one person is actually controlling the game. In contrast, a commercial from Japan shows a woman playing one of the minigames alone—talking to herself and laughing about her performance.[112] Another Japanese spot features no people; instead, an unseen announcer explains different aspects of the updated game while shots of the balance board and gameplay are shown.[113] These commercials offer a much more nuanced perspective on the possibilities of play, pleasure, and sociability—and reflect a more diverse group of potential players.

CONCLUSION

In this chapter, I have examined the ways in which the Wii, and specifically *Wii Fit*, disciplines the body and reinscribes stereotypical notions of gender, play, pleasure, and embodiment. I explained how *Wii Fit* conforms to and challenges Jesper Juul's casual game typology and explored how it fits into the exergaming genre. I also interrogated the ways in which the Wii console configures the domestic sphere, and encourages pleasure through sociable gameplay.

In its choice to emphasize intuitive, physical gameplay over photo-realistic graphics, Nintendo created a console that appeals to many female and casual players. In addition, other aspects of the Wii, such as the androgynous, cartoony Miis each player can create and share, reinforce the notion that these sorts of players prefer immersive and social gameplay over visual richness and competition. While the Wii (and *Wii Fit*) are likely to bring new players into the console gaming world, it is unclear if they will be equally interested in extending their play beyond exergames.

The exergaming genre is ripe for further explanation. In particular, further research into the social practices surrounding these games and whether they encourage long-term fitness changes is critical. Especially important is determining how to encourage individuals to engage in healthy behaviors while not patronizing or chastising them. It would also be worthwhile to know how individuals view these games—are they pleasurable, work, or some of both? And for game designers, continued understanding of the ways in which players make sense of and integrate these games into their lives could prove invaluable.

NOTES

1. Tristan Donovan, *Replay: The History of Video Games* (East Sussex, UK: Yellow Ant, 2010).

2. Jesper Juul, *A Casual Revolution: Reinventing Video Games and Their Players* (Cambridge, MA: MIT Press, 2010); "Casual Games Association," *Casual Games Market Report 2007: Business and Art of Games for Everyone* (Casual Games Association, 2007), accessed at http://www.casualgamesassociation.org/pdf/2007_CasualGamesMarketReport.pdf.

3. Katie Conboy, Nadia Medina, and Sarah Stanbury, "Introduction." In *Writing on the Body: Female Embodiment and Feminist Theory*, ed. by Katie Conboy, Nadia Medina, and Sarah Stanbury (New York: Columbia University Press, 1997), 1.

4. Chris Castiglione, "Mci 1997 Anthem Campaign" (Vimeo, LLC, December 6, 2008), accessed at http://www.vimeo.com/2445340. For an excellent discussion of this commercial, see Lisa Nakamura, "'Where Do You Want to Go Today?': Cybernetic Tourism, the Internet, and Transnationality." In *Race in Cyberspace*, ed. by Beth E. Kolko, Lisa Nakamura, and Gilbert B. Rodman (New York: Routledge, 2000), 15–26.

5. William Gibson, *Neuromancer* (New York: Ace, 1984).

6. Neal Stephenson, *Snow Crash* (New York: Bantam Dell, 1993).

7. Examples of this utopian vision of technology include Sherry Turkle, *Life on the Screen: Identity in the Age of the Internet* (New York: Simon and Schuster, 1995) and Howard Rheingold, *The Virtual Community: Homesteading on the Electronic Frontier* (Reading, MA: Addison-Wesley, 1993). An excellent discussion that traces the origins of this utopian discourse is found in David Bell, *An Introduction to Cyberculture* (London: Routledge, 2001).

8. Donna J. Haraway, "A Cyborg Mainfesto: Science, Technology, and Socialist-Feminism in the Late Twentieth Century." In *Simians, Cyborgs and Women: The Reinvention of Nature*, ed. by Donna Jeanne Haraway (New York: Routledge, 1991), 149–182.

9. James Cameron, *The Terminator* (USA: Orion Pictures Corporation, 1984); James Cameron, *Terminator 2: Judgment Day* (USA: TriStar Pictures, 1991).

10. Paul Verhoveven, *Robocop* (USA: Orion Pictures Corporation, 1987).

11. Paul Verhoeven, *Total Recall* (USA: TriStar Pictures, 1990).

12. The Wachowski Brothers, "The Matrix," (USA: Warner Bros. Pictures, 1999).

13. David Cronenberg, *Existenz* (USA: Dimension Films, 1999).

14. Claudia Springer, "The Pleasure of the Interface." In *Cybersexualities: A Reader on Feminist Theory, Cyborgs and Cyberspace*, ed. by Jenny Wolmark (Edinburgh: Edinburgh University Press, 1999), 34.

15. Lisa Nakamura, *Cybertypes: Race, Ethnicity, and Identity on the Internet* (New York: Routledge, 2002).

16. Mimi Ito, "Kids and Simulation Games: Subject Formation through Human-Machine Interaction" (paper presented at the annual meeting of the Society for the Social Studies of Science, Tucson, AZ, October, 1997).

17. Clay Spinuzzi, *Tracing Genres through Organizations: A Sociocultural Approach to Information Design* (Cambridge, MA: MIT Press, 2003).

18. See Edward Castronova, *Synthetic Worlds: The Business and Culture of Online Games* (Chicago: University of Chicago Press, 2005), 4–9.

19. Mark Stevens Meadows, *I, Avatar: The Culture and Consequences of Having a Second Life* (Berkeley, CA: New Riders, 2008), 13.

20. Adriano D'Aloia, "Adamant Bodies. The Avatar-Body and the Problem of Autoempathy," *E|C 3*, 5(2009): 51.

21. T. L. Taylor, *Play between Worlds: Exploring Online Game Culture* (Cambridge, MA: MIT Press, 2006), 96.

22. Teresa de Laurentis, *Technologies of Gender: Essays on Theory, Film, and Fiction* (Bloomington: Indiana University Press, 1987), 5. Emphasis in original.

23. Judith Butler, *Bodies That Matter: On the Discursive Limits of "Sex"* (New York: Routledge, 1993), 2–3.

24. Trevor Pinch and Wiebe E. Bijker, "The Social Construction of Facts and Artifacts: Or How the Sociology of Science and the Sociology of Technology Might Benefit Each Other," *Social Studies of Science 14* (1984): 399–441.

25. Nina E. Lerman, Arwen Palmer Mohun, and Ruth Oldenziel, "The Shoulders We Stand on and the View from Here: Historiography and Directions for Research," *Technology and Culture 38* 1(1997): 9–30.

26. Stephen Strauss, "Video-Game Firms Are All Set to Zap Women's Market" (*The Globe and Mail*, August 5, 1982); Anne McCarroll, "'Hey, Some Kid's Mom Is Playing'" (*Christian Science Monitor*, August 30, 1982).

27. Justine Cassell and Henry Jenkins, eds., *From Barbie to Mortal Kombat: Gender and Computer Games* (Cambridge, MA: MIT Press, 1998).

28. Henry Jenkins and Justine Cassell, "From Quake Grrls to Desperate Housewives: A Decade of Gender and Computer Games." In *Beyond Barbie & Mortal Kombat: New Perspectives on Gender and Gaming*, ed. by Yasmin B. Kafai et al. (Cambridge, MA: MIT Press, 2008), 7–9.

29. Jennifer Glos and Shari Goldin, "An Interview with Brenda Laurel (Purple Moon)." In *From Barbie to Mortal Kombat: Gender and Computer Games,* ed. by Justine Cassell and Henry Jenkins (Cambridge, MA: MIT Press, 1998), 118–135.

30. Kaveri Subrahmanyam and Patricia M. Greenfield, "Computer Games for Girls: What Makes Them Play?" In *From Barbie to Mortal Kombat: Gender and Computer Games*, ed. by Justine Cassell and Henry Jenkins (Cambridge, MA: MIT Press, 1998), 46–71.

31. Henry Jenkins, "Voices from the Combat Zone: Game Grrlz Talk Back." In *From Barbie to Mortal Kombat: Gender and Computer Games*, ed. by Justine Cassell and Henry Jenkins (Cambridge, MA: MIT Press, 1998), 328–341.

32. Nick Dyer-Witheford and Greig de Peuter, *Games of Empire: Global Capitalism and Video Games* (Minneapolis: University of Minnesota Press, 2009).

33. Mark D. Griffiths, Mark N. Davies, and Darren Chapell, "Breaking the Stereotype: The Case of Online Gaming," *Cyberpsychology & Behavior 6*, 1(2003): 81–91.

34. Cornelia Brunner, Dorothy Bennett, and Margaret Honey, "Girl Games and Technological Desire." In *From Barbie to Mortal Kombat: Gender and Computer Games*, ed. by Justine Cassell and Henry Jenkins (Cambridge, MA: MIT Press, 1998), 72–87.

35. Dorothy Bennett, Cornelia Brunner, and Margaret Honey, "Gender and Technology: Designing for Diversity" (New York: Education Development Center, Inc. Center for Children and Technology, 1999), 5.

36. Taylor, *Play between Worlds.*

37. Pam Royse et al., "Women and Games: Technologies of the Gendered Self," *New Media & Society 9*, 4(2007): 556.

38. Taylor, *Play between Worlds,* 94–100.

39. Elisabeth Hayes, "Women, Video Gaming & Learning: Beyond Stereotypes," *Tech Trends 49*, 5(2005): 24.

40. Jesse Schell, *The Art of Game Design: A Book of Lenses* (Burlington, MA: Morgan Kaufmann, 2008).

41. Jesse Schell, *The Art of Game Design*, 104–105.

42. Jesse Schell, *The Art of Game Design*, 103.

43. For examples of these kinds of heteronormative, male-focused representations, see Brad Gallaway, "Mass Effect 2 and the Lack of Homosexuality in Space" (GameCritics.com, February 12, 2010), accessed at http://www.gamecritics.com/brad-gallaway/mass-effect-2–and-the-lack-of-homosexuality-in-space; and Alex Raymond, "Women Aren't Vending Machines: How Video Games Perpetuate the Commodity Model of Sex" (GameCritics.com, August 9, 2009), accessed at http://www.gamecritics.com/alex-raymond/women-arent-vending-machines-how-video-games-perpetuate-the-commodity-model-of-sex.

44. "Nintendo Annual Report" (Nintendo Co., Ltd., 2009).

45. Matt Cabral, "Play Magazine Online: Review–Ghostbusters," accessed at http://play-magazine.com/?fuseaction=SiteMain.Content&contentid=1744.

46. Ben Kuchera, "Nintendo Wii: The Ars Technica, Here Wii Go" (Ars Technica, 2011), accessed at http://arstechnica.com/hardware/reviews/2006/11/wii.ars.

47. "Wii Sports" (Nintendo, 2006).

48. Mihaly Csikszentmihalyi, *Flow: The Psychology of Optimal Experience* (New York: Harper & Row, 1990).

49. Juul, *A Casual Revolution*, 103–107.

50. Steven E. Jones, *The Meaning of Video Games: Gaming and Textual Strategies* (New York: Routledge, 2008), 134.

51. Juul, *A Casual Revolution*, 114–115.

52. Jeff Bakalar, "Review: Nintendo Wii (Original, Wii Sports Bundle)" (CBS Interactive, 2011), accessed at http://reviews.cnet.com/consoles/nintendo-wii/4505–10109_7–31355104.html#cnetReview; Kuchera,"Nintendo Wii."

53. Daemon Hatfield, "GDC 2007: 'The Wii is a Piece of $#&%!'" (IGN Entertainment, October 31, 2008), accessed at http://wii.ign.com/articles/771/771051p1.html.

54. Kris Graft, "Games at-a-Glance: Wii Quality" (*Edge Online*, 2009), accessed at http://www.edge-online.com/features/games-at-a-glance-wii-quality.

55. An example of the Mii creation screen can be found at http://www.miiware.com/alt-index2.php?mainspot=MiiMaker.php.

56. For example, in "MarioKart Wii," players can unlock racing suit outfits for their Miis. See "Mario Kart Wii—Super Mario Wiki," http://www.mariowiki.com/Mario_Kart_Wii.

57. Nintendo, "Part 2 at Nintendo: Wii: What Is Wii?: Iwata Asks: Volume 1" (Nintendo, 2011), accessed at http://www.nintendo.com/wii/what/iwataasks/volume-1/part-2.

58. Jones, *The Meaning of Video Games.*

59. Taylor, *Play between Worlds.*

60. Bonnie Ruberg, "Playing Dirty: Women on the Wii," *joystiq,* November 16, 2006, accessed at http://www.joystiq.com/2006/11/16/playing-dirty-women-on-the-wii/.

61. Bernadette Flynn, "Geography of the Digital Hearth," *Information, Communication & Society 6,* 4(2003): 551–576.

62. Flynn, "Geography of the Digital Hearth," 557–558.

63. Amy Voida and Saul Greenberg, "Wii All Play: The Console Game as a Computational Meeting Place" (paper presented at the annual meetings of the CHI 2009, Boston, MA, April, 2009).

64. Allison Sall and Rebecca E. Grinter, "Let's Get Physical! In, Out and Around the Gaming Circle of Physical Gaming at Home," *Computer Supported Cooperative Work 16,* 1/2(2007): 199–229.

65. Helen Thornham, "Claiming a Stake in the Videogame: What Grown-Ups Say to Rationalize and Normalize Gaming," *Convergence 15,* 2 (2009): 141–159.

66. Helen Thornham, "Claiming a Stake in the Videogame," 148.

67. Helen Thornham, "'It's a Boy Thing': Gaming, Gender, and Geeks," *Feminist Media Studies 8,* 2(2008): 127–142.

68. Thornham, "'It's a Boy Thing.'"

69. Voida and Greenberg, "Wii All Play."

70. Gareth R. Schott and Siobhan Thomas, "The Impact of Nintendo's 'For Men' Advertising Campaign on a Potential Female Market," *Eludamos 2,* 1(2008): 47.

71. Barbara O'Connor and Elisabeth Klaus, "Pleasure and Meaningful Discourse: An Overview of Research Issues," *International Journal of Cultural Studies 3,* 3(2000): 369–387.

72. John Fiske, "TV: Re-Situating the Popular in the People," *Continuum: The Australian Journal of Media & Culture 1,* 2(1987): 56–66.

73. O'Connor and Klaus, "Pleasure and Meaningful Discourse," 377–378.

74. Aphra Kerr, Julian Kücklich, and Pat Brereton, "New Media—New Pleasures?" *International Journal of Cultural Studies 9,* 1(2006): 63–82.

75. "Wii Fit" (Nintendo, 2008).

76. Andre Hilsendeger et al., "Navigation in Virtual Reality with the Wii Balance Board," *6th Workshop on Virtual and Augmented Reality* (2009), accessed at http://www.techfak.uni-bielefeld.de/ags/wbski/hiwis/ahilsend/files/naviga-tion_in_virtual_reality_with_the_wii_balance_board.pdf.

77. Anonymous, "Wii Balance Board Games List (Wii Fit Games)" (Blogger, N.d.), accessed at http://www.balanceboardblog.com/2008/06/guide-every-wii-balance-board-game.html.

78. A. J. Glasser, "Wii Fit Plus Review: Now I'm a Believer" (Kotaku.com, October 6, 2009), accessed at http://kotaku.com/5375517/wii-fit-plus-review-now-im-a-believer; Chris Kohler, "Hands On: New Wii Fit Plus Workout's Custom Routines" (Wired.com, October 5, 2009), accessed at http://www.wired.com/gamelife/2009/10/wii-fit-plus/.

79. "Casual Games Association," *Casual Games Market Report 2007: Business and Art of Games for Everyone* (Casual Games Association, 2007), accessed at http:// www.casualgamesassociation.org/pdf/2007_CasualGamesMarketReport.pdf.

80. Juul, *A Casual Revolution*, 50.

81. Debra A. Lieberman, "Dance Games and Other Exergames: What the Research Says" (Unpublished manuscript, 2006), accessed at http://www.comm.ucsb.edu/faculty/lieberman/exergames.htm.

82. Ian Bogost, "The Rhetoric of Exergaming" (paper presented at the annual meetings of the Digital Arts & Cultures, Copenhagen, Denmark, December, 2005).

83. Bogost, "The Rhetoric of Exergaming."

84. Sunny Consolvo et al., "Design Requirements for Technologies that Encourage Physical Activity" (paper presented at the annual meetings of the CHI, Montreal, Canada, April 2006).

85. See Anne Underwood, "The Wii Fit Workout: Can a Videogame Help You Lose Weight?" (*Newsweek*, May 20, 2008), accessed at http://www.newsweek.com/2008/05/20/the-wii-fit-workout.html; and Lee Graves et al., "Comparison of Energy Expenditure in Adolescents When Playing New Generation and Sedentary Computer Games: Cross Sectional Study," *British Medical Journal, 335* (2007): 1282–1284. Accessed at http://www.bmj.com/cgi/content/full/335/7633/1282.

86. "The President's Challenge—Adult Fitness Test," accessed at http:// www.adultfitnesstest.org/adultFitnessTestLanding.aspx.

87. Stew Smith, "Military Fitness: Army Basic Training Pft" (Military.Com, 2011), accessed at http://www.military.com/military-fitness/army-fitness-requirements/army-basic-training-pft.

88. Dianne Neumark-Sztainer et al., "Self-Weighing in Adolescents: Helpful or Harmful? Longitudinal Associations with Body Weight Changes and Disordered Eating," *Journal of Adolescent Health 39*, 6(2006): 811–818.

89. "Centers for Disease Control and Prevention," "Healthy Weight – It's Not a Diet, It's a Lifestyle!: About BMI for Adults" (Centers for Disease Control and Prevention, N.d.), accessed at http://www.cdc.gov/healthyweight/assessing/bmi/adult_bmi/index.html.

90. "Centers for Disease Control and Prevention," "Healthy Weight—It's Not a Diet."

91. Peta Bee, "The BMI Myth" (*The Guardian*, November 28, 2006), accessed at http:// www.guardian.co.uk/lifeandstyle/2006/nov/28/healthandwellbeing.health1.

92. Kate Harding, "Why BMI Is a Crock, in Pictures" (Shakesville, September 29, 2007), accessed at http://shakespearessister.blogspot.com/2007/09/why-bmi-is-crock-in-pictures.html.

93. Kate Harding, "Illustrated BMI Categories" (Yahoo! Inc., 2011), accessed at http:// www.flickr.com/photos/77367764@N00/sets/72157602199008819/.

94. The American Council on Exercise, "Test Results Reported on Fitness Benefits of Nintendo's Wii Fit and Pc-Based Exergame, Dancetown" (The American Council on Exercise, November 11, 2009), accessed at http://www.acefitness.org/pressroom/442/test-results-reported-on-fitness-benefits-of/.

95. The American Council on Exercise, "Test Results Reported on Fitness Benefits of Nintendo's Wii Fit and Pc-Based Exergame, Dancetown."

96. Michel Foucault, "Docile Bodies." In *The Foucault Reader*, ed. by Paul Rabinow (New York: Pantheon Books, 1984), 179–187.

97. Roland Marchand, *Advertising the American Dream: Making Way for Modernity, 1920–1940* (Berkeley: University of California Press, 1985).

98. Mojca Pajnik and Petra Lesjak-Tušek, "Observing Discourses of Advertising: Mobitel's Interpellation of Potential Consumers," *Journal of Communication Inquiry 26*, 3(2002): 281.

99. Louis Althusser, "Ideology and Ideological State Apparatuses (Notes Towards an Investigation)." In *Lenin and Philosophy and Other Essays*, ed. by Louis Althusser (New York: Monthly Review Press, 1971), 127–186.

100. Stuart Hall, "Encoding/Decoding." In *Media and Cultural Studies: Keyworks*, ed. by Meenakshi Gigi Durham and Douglas M. Kellner (Malden, MA: Blackwell, 2001), 163–173.

101. Nelly Oudshoorn and Trevor Pinch, "How Users and Non-Users Matter." In *How Users Matter: The Co-Construction of Users and Technology*, ed. by Nelly Oudshoorn and Trevor Pinch (Cambridge, MA: MIT Press, 2005), 67–68.

102. Erica Scharrer, "Virtual Violence: Gender and Aggression in Video Game Advertisements," *Mass Communication & Society 7*, 4(2004): 393–412.

103. Bart Simon, "Wii Are out of Control: Bodies, Game Screens and the Production of Gestural Excess" (Social Science Research Network, 2009), accessed at http://ssrn.com/paper=1354043.

104. In addition to advertising the product, Nintendo also launched a substantial public relations effort, which included giving out $15 gift cards to willing individuals who would exercise using *Wii Fit Plus* in the parking lot of several Target stores. See Alexander Sliwinski, "Nintendo and Target Teaming up for *Wii Fit Plus* Promo—Receive $15 Gift Card" (Joystiq.com, January 14, 2010), accessed at http://nintendo.joystiq.com/2010/01/14/nintendo-and-target-teaming-up-for-wii-fit-plus-promo-receive/.

105. Anonymous, "Nintendo Wii Fit: 'Working Mom'" (Motionographer, n.d.), accessed at http://motionographer.com/theater/nintendo-wii-fit-working-mom/.

106. Restricted access available at http://www.youtube.com/watch?v=kYys0QvAzNE.

107. Thornham, "Claiming a Stake in the Videogame."

108. "Anonymous," "Wii Fit Plus" (GAMERPRESSURE, 2011), accessed at http://gameads.gamepressure.com/tv_game_commercial.asp?ID=11000.

109. "Anonymous," "Wii Fit Plus" (GAMERPRESSURE, 2011), accessed at http://gameads.gamepressure.com/tv_game_commercial.asp?ID=10998; and "Anonymous," "Wii Fit Plus," (GAMERPRESSURE, 2011), accessed at http://gameads.gamepressure.com/tv_game_commercial.asp?ID=10997.

110. "Anonymous," "Wii Fit Plus" (GAMERPRESSURE, 2011), accessed at http://gameads.gamepressure.com/tv_game_commercial.asp?ID=10999.

111. majkeld2, "Wii Fit Plus Boxing Commercial Nintendo Wii for Korea Asia Gameads Fit" (YouTube.com, January 30, 2010), accessed at http://www.youtube.com/watch?v=NgP7ZaXpaQ4.

112. "Anonymous," "Wii Fit Plus" (GAMERPRESSURE, 2011), accessed at http://gameads.gamepressure.com/tv_game_commercial.asp?ID=10617.

113. NintenDaanNC, "[Minna no NC] Wii Fit Plus—Commercial" (YouTube.com, September 2, 2009), accessed at http://www.youtube.com/watch?v=ZXTU6tRliTU.

Chapter Eight

Sincere Fictions of Whiteness in Virtual Worlds

How Fantasy Massively Multiplayer Online Games Perpetuate Color-blind, White Supremacist Ideology

Joel Ritsema and Bhoomi K. Thakore

The rise in popularity of Massively Multiplayer Online Role Playing Games (MMORPGs) has presented questions that confound traditional understandings of social life. This has ignited a new body of literature in which scholars seek to expand our knowledge of social interaction in virtual spaces.[1] Among the important themes discussed in contemporary scholarship on MMORPGs, scholars note that these virtual worlds may allow for some renegotiation of the social boundaries that organize and stratify society at large. Researchers have been enthusiastic about their "discovery" of MMORPGs as vibrant third places[2] in which deep personal relationships are fostered. These researchers assert that MMORPGs construct new virtual worlds in which individuals engage in collaborative play, share experiences and rewards, and are socialized into communities of gamers.[3] In this same vein, MMORPGs (and cyberspace in general) are presented in the literature as spaces of opportunity for strategic identity construction in which individuals have unprecedented control to explore and present new identities, defying the trappings of race, class, and gendered stereotypes and the subsequent set of corresponding social interactions within these virtual spaces.[4]

The newness and emergent nature of the MMORPG as a medium for social interaction seems to suggest that the rules of engagement within these spaces are largely unwritten. Virtual places could offer limited freedom from the various physical, social, political, and economic constraints which function to separate and stratify society as we know it. Indeed, this idea seems to be shared widely by scholars and members of the gaming community alike, and many have been enthusiastic in embracing the emancipatory nature of virtual worlds. As Margaret Wertheim notes in her article so aptly titled "The Pearly Gates of Cyberspace,"

> Just as early Christians envisaged heaven as an idealized realm beyond the chaos and decay of the material world . . . so too, in this time of social and environmental disintegration, today's proselytizers of cyberspace proffer their domain as an ideal "above" and "beyond" the problems of the material world. While early Christians promulgated heaven as a realm in which the human soul would be freed from the frailties and failings of the flesh, so today's champions of cyberspace hail it as a place where the self will be freed from the limitations of physical embodiment. [5]

Despite widespread interest in the vast emancipatory potential of virtual worlds, it is important to guard against adopting a romantic perspective that inhibits one from clearly seeing the ways in which social inequalities penetrate the virtual worlds of MMORPGs. In proffering the egalitarian and emancipatory nature of virtual worlds, optimistic scholars often overlook how gamers who enter these worlds, and game designers that set the stage of play, have been markedly unsuccessful in escaping the social realities of inequality that are so deeply imbedded in the "outside world."

We therefore wish to advocate for scholarship on MMORPGs that focus on how systems of domination related to race, gender, sexuality, and social class are perpetuated in these games. We argue that insofar as they exist in a society characterized by hierarchically arranged categories of race, class, gender, and sexuality, these virtual worlds ultimately assume a normative framework that privileges white heterosexual masculinity and subjugates or represses those who do not fit within it.

In this chapter, we focus specifically on issues of racial inequality in many MMORPGs by addressing the under-analyzed racialization of virtual spaces and the ways in which these games reflect and reproduce the racialized social systems existing in U.S. society. Emphasizing the ways in which racial tropes and colonial representations of certain groups of people are allegorically represented in the virtual worlds of many MMORPGs, we argue that these worlds offer "sincere fictions of whiteness" that function to maintain a white dominant frame, even amidst claims of "racelessness" or color-blindness. [6] We draw from Feagin's [7] notion of sincere fictions of the white self to explore the racialized systems of meaning and patterns of interaction evident in these virtual spaces. In so doing, we argue that while the virtual worlds in MMORPGs may indeed provide new opportunities for identity construction and rich social interaction, these opportunities are not the same for everyone; the racialized nature of these virtual worlds produces social contexts in which some players are limited and excluded from fully exploring these opportunities.

VIRTUAL WORLDS AND SOCIAL IDENTITY

As previously noted, virtual worlds have been presented in the literature as strategic spaces of identity construction and presentation, producing opportunities for more egalitarian social interaction. Researchers who contend that MMORPGs provide emergent spaces for rich social life and community-building point out that these spaces are perhaps less constraining than traditional spaces for interaction because they in some ways eliminate important physical and social boundaries rooted in "face-to-face" interaction.

As Goffmann[8] notes, traditional methods of presenting an identity to others in face-to-face (as opposed to virtual) interaction rely on "sign vehicles"—clothing, body language, gestures, etc.—that carry shared symbolic meaning with the intended audience. These "sign vehicles" are often used spontaneously and are thus outside the individual's sphere of control. Consequently, presenting an identity can be a precarious and delicate process, loaded with ever-present risks of accidents, embarrassment, and the misinterpretation of messages by the presenter's intended audience.[9] Researchers contend that computer mediated communication eliminates many of these risks in self-presentation and allows individuals greater control over the stages on which they are displayed. As such, virtual places provide individuals with unique opportunities to operate strategically in constructing an identity and presenting it to others. Scholars[10] assert virtual spaces provide individuals with such control in impression management that one can even project different ages, races, gender identities, and other personal characteristics that would otherwise bind them to certain socially constructed scripts and expectations. As Jones explicitly argues in his text on virtual culture, "The internet eliminates the interpersonal identification and judgment processes by which we normally evaluate each other in face-to-face interaction."[11]

With regard to race, if we adhere uncritically to this perspective (as people often do), we could easily conclude that issues of race no longer exist, let alone matter, in MMORPGs' virtual worlds, as hierarchical arrangements based upon socially constructed notions of race that pervade society are overcome by new opportunities for identity (re)construction. Indeed, it may seem at first glance that MMORPGs "free" individuals from race and racism, allowing them to pursue social goals and relationships unfettered by social constraints. We could argue that race as it is understood in society at large has no real meaning in the virtual world.

Indeed, the belief that virtual worlds are "race-less" spaces enters the discourse on MMORPGs in everyday interactions in and around these games. Higgin[12] reveals one example in his article on race and MMORPGs. It involves a popular home-made video posted online depicting a group of individuals engaged in collaborative play in the popular game *World of Warcraft*. The video, which portrays a dark-skinned character carelessly charging into battle before his teammates are ready, draws heavily on historically-rooted stereotypes of the black American as comedic buffoon. As his teammates admonish him, the player replies, "at least I have chicken." Higgin notes that in an online discussion about this video, one individual highlights these racial elements and is immediately sanctioned by several others. One person replies, "I'm confused. He's a character in a game. He doesn't have a race. . . . Maybe the person who is racist is you."[13]

In another example, during an online discussion of Lastowka's article[14] on "cultural borrowing" in the making of *World of Warcraft*, one respondent argues, "It[']s a fucking Game, no racism, or Political culture or any other mambo jambo you are talking about. Plz, if you want to analyze things, do it in real things." Interactions like these reveal how racism operates within these virtual places, but also how it can so easily (and so often) be disregarded as occurring in "just a game," with no real link to society at large, and no real social consequences.

Despite what the individuals in the above examples argue, race remains a salient part of both the material and social worlds of MMORPGs.[15] Not only do structures in the virtual "worlds" of MMORPGs regularly display racialized imagery, but the storylines themselves operate in ways that maintain a dominant white, Eurocentric perspective. The virtual spaces constructed by these game designers and players thus reflect, and indeed reinforce, issues of race that are woven into the social fabric of U.S. society at large.

MMORPGS' RACIST HISTORY: RACIAL ELEMENTS IN HIGH FANTASY LITERATURE

One salient feature of many fantasy-based MMORPGs is related to how they reproduce racialized colonial images and ideology. Fantasy games like *Lord of the Rings Online*, *World of Warcraft*, and others borrow heavily from the profoundly influential works of "high fantasy" created by J.R.R. Tolkien and C. S. Lewis, whose works have also drawn criticism for raising racist and colonial imagery. It is beyond the focus of this paper to detail each of these criticisms, but they are relevant to this work insofar as they raise issues of race in these authors' works that are similar to those found in many MMORPGs. Tolkien's *The Lord of the Rings* series and *The Hobbit* present a narrative of massive struggle between good and evil, and the characteristics of those on either end of this moral spectrum are often hyper-racialized. Tolkien assigns his fantastical races both physical and geographical characteristics. Scholars[16] have critiqued Tolkien for his portrayal of the "orcs" and "mercenary humans" from the South and East as dark, "swarthy-skinned," "slant-eyed," and evil, while the human and "elvish" heroes from the West as noble, "fair," "white," and "yellow-haired." The weapons, armor, and vehicles used by those from the South and East have also been criticized for referencing the ancient Mongol and Saracen armies, while the armor, clothing, and castles of hobbits and humans draw extensively from medieval Western Europe.[17] Likewise, C. S. Lewis' *Chronicles of Narnia* has drawn fire for rather unambiguous Arab-bashing (the dwarves refer to the evil, turban-wearing Calormenes as "darkies" throughout *The Last Battle*).[18]

Issues of racial allegory continued as these fantasy works were converted into motion pictures. The *Lord of the Rings* films released in 2001–2003 based on Tolkien's trilogy reflect, and indeed further intensify, the racial allegory found in the original texts. As Kim[19] points out in her article on the series, the racial coding drawn from the novels is particularly emphasized in the films, with the heroic humans and hobbits depicted as Western and Northern European, and the evil forces composed of dark-skinned, brutish orcs and dark-skinned, savage, dreadlocked humans wearing turbans and riding giant elephants.

Considerable debate has ensued regarding these issues of race in Tolkien's novels and in the *Lord of the Ring* films, and much of it has become entangled in discussions of whether Tolkien or C. S. Lewis were racist, or whether the filmmakers are racist, etc. Because these arguments focus on the motivation or intention of the author and filmmakers, they fail to acknowledge how issues of racism, which are embedded in the fabric of Western society, penetrate these mediums and reflect dominant racial ideologies rather than individual intent. As we note in further detail below, sincere fictions are *sincere* because they may be expressed with the most benign of intentions, denying or lacking an awareness of how they normalize and privilege certain racial perspectives and suppress alternatives. The content in this high fantasy literature and films can—and should—be examined independently from their creators' stated (or unstated) intentions to gain a broader sense of how racial ideologies operate in contemporary forms of mass media.

We therefore seek to move beyond the superficial discussions of race in fantasy media that focus on individual motives and actions to explore how these media function, intentionally or not, to reflect and reinforce the racial order of contemporary society. We adhere to the perspective that sees the concept of race itself as a fantasy, or a fiction—as socially constructed classifications of people, ordered in a social hierarchy extending undeserved privileges and benefits to some, while penalizing and oppressing others.

While race, racism, and its less visible structural presentation in everyday life has psychological, visceral, and material consequences, race as a contemporary phenomenon is not a static identity based upon biological characteristics. The concept of race is fluid, continuously in development and renegotiated across varied terrains of social life, while its direction is subsequently shaped by the ideology and social realities of dominant groups.[20] How race and race relations are constructed by these dominant groups is displayed in various mass media outlets ranging from books to films, and of course, to the virtual worlds of MMORPGs.

RACIAL ELEMENTS IN MMORPGS

Popular fantasy MMORPGs often lift, illuminate, and expand upon the same racial tropes that exist in Tolkien's work described above. Notions of good and evil in the *Lord of the Rings Online* are consistent with the books and films, and the game provides players with opportunities to enter the fantastical world of "Middle Earth," provided they adopt a white appearance. Character choices in *Lord of the Rings Online* are limited to white phenotypes, and main characters and allies also exclusively appear to be of European descent. It is only once a player reaches a certain level with a white avatar that they are able to create a monster avatar if they desire to align themselves with "evil" forces, as choosing a non-white avatar is simply not an option in *Lord of the Rings Online.*

The same is true for other fantasy-based MMORPGs that provide very limited, if any, opportunities for engaging in the storyline as a non-white character. As Higgin[21] notes, black heroes remain conspicuously absent in fantasy MMORPGS, and this whitewashing of fantasy game protagonists may actually have increased as new games are released. Higgin argues that because these games are based so studiously upon a particular body of literature which sought to "create a form of western European mythology," they necessarily embrace the dominant white perspective of that literature which reveres mythical white heroes and draws upon stereotypes of racial minorities as reference points for constructing evil, barbaric enemies.

In *World of Warcraft*, the most popular MMORPG created to date, different fantastical races are divided into two warring factions, the Alliance and the Horde, and racial allegory is as evident in this game as it is in the Tolkien novels and films. "The Alliance" is composed of humans, dwarves, gnomes, and night elves; "the Horde" is composed of orcs, trolls, tauren, and the undead. The characteristics of these races draw from certain real-world racial and cultural ideas. While this racial allegory has been discussed extensively in online forums, little of this discussion has been addressed at length in scholarly work and deserves to be reemphasized here. The Orcs, who the game designers note[22] are "commonly believed to be brutal and mindless," are beastly, aggressive, and apelike, with tusks protruding from their mouths. Their home city and surrounding areas are obvious tropes of Africa, complete with dry river beds, savannahs, giraffes, and zebras. The trolls, who "populate the islands of the south seas," are "renowned for their cruelty and dark mysticism," and are described as "wily," "barbarous and superstitious."[23] Trolls also display huge tusks and their speech contains obvious references to stereotypical West Indies accents, complete with occasional "Hey, mon!" expressions. The Tauren race is a strange combination of bovine and human; they walk upright, but have hooves for feet, horns on their heads, and fur-covered bodies. The game website describes the Tauren as "huge, bestial creatures" who "cultivate a quiet, tribal society."[24] The Tauren reside in villages of teepees, lodges, and totem poles on the "grassy, open barrens of central Kalimdor" that seem clearly representative of the Great Plains of North America. Finally, the game website describes the Undead as "dark warriors" with "dark aims" who have "entered into an alliance of convenience with the *primitive, brutish* races of the Horde."[25]

In stark contrast to the Horde, the races of the Alliance clearly draw from Western European cultural and racial imagery. The humans, virtually all of whom are white, are described on the game website as "resilient," "heroic," and "fierce warriors" who "stand resolute in their charge to maintain the honor and might of humanity in an ever-darkening world."[26] Their home is the city of "Stormwind," complete with Western-style castles, spires, and houses, surrounded by lush temperate forests and farmland. The dwarves, described as "robust humanoids" with "unmatched courage and valor,"[27] live in a large city with stone Western-European style towers, carved out of a snow- and pine-covered mountain range. The dwarves speak with Scottish accents, again emphasizing a Western European identity. The gnomes, which are described as an "eccentric, often-brilliant" race who "spend their time devising strategies and weapons that will help . . . build a brighter future for their people,"[28] share both the snowy mountains and white phenotypes of the dwarves. The night elf race is perhaps the most ambiguous in its connection to a specific "real-world" racial category, but their lack of whiteness seems to strain their relationship even in the fantasy *World of Warcraft*; the game website notes that the relationship between the night elves and the other races of the Alliance is often characterized by distrust.

A number of issues are raised by these racial allegories in *World of Warcraft*. First, a perpetual conflict exists between characters that represent white Western society, and characters based upon marginalized and colonized real-world races. At face value,[29] the Alliance races are the obvious "good guys" with superior culture, intellect, and morality, and the Horde are the obvious "bad guys," vicious, primitive, and superstitious. This stark contrast reflects and adds to the representation of moral and intellectual superiority on the basis of race that is seen in other fantasy media. While much speculation has been made about whether the creators of *World of Warcraft* intend the Horde or Alliance to be the "good guys," we again must move beyond conjectures about intention to examine the larger social forces at play that extend beyond game designers' sphere of influence. Regardless of who is intended to be "good" and "bad," it is important to note that within this fantasy world, it is perpetual conflict—not reconciliation—that drives relations between factions. Furthermore, due to the emergent nature of MMORPGs, much of what is defined as "good" and "bad" can be interpreted and shaped by those who play the game. But this interpretation does not occur in social isolation—it is guided by the interfacing of the racialized imagery of the game with players' life experiences within a racialized social system.

Second, not only do the Horde races draw from racial stereotypes in terms of dress, architecture, and environment, but their tusks, horns, fur, and hooves also function to identify them as other-than-human. In this way, as Higgin[30] has noted, not only do white races of the Alliance exclusively posses the qualities of "heroism," "unmatched courage and valor," and "brilliance," but they also claim exclusive ownership of what it means to be fully human. This creates a conspicuous absence of non-white human characters in *World of Warcraft*, carving out interpretive space for the construction of sincere fictions which define whiteness, and only whiteness, as morally upright and completely human.

SINCERE FICTIONS

As social, self-aware beings, we construct our identities through reflexive processes of accepting or rejecting what we see in others, or what others tell us about ourselves. However, some members of society possess more agency in determining how they are defined by others. In the United States today, these discrepancies in power to define self and others are based in no small part on socially constructed and hierarchically arranged categories of race. Sincere fictions of the white self are strategically constructed ideas and images of what it means to be white. Whiteness in this sense is not simply a physical characteristic of individuals, but represents a dominant group holding the power not only to order the racial social hierarchy, but also to conceal this dominance in ways that make it appear as benign, normative, and natural.

The notion of the sincere fiction is rooted in the theories of a number of cultural sociologists. Mauss[31] introduces this when he explores the act of gift-giving as a social action that usually obliges reciprocation. When one gives a gift to another, Mauss observes, a social bond is created between giver and recipient which generally dictates that the recipient must return the favor. However, individuals regularly pretend that gifts are freely and voluntarily given, and conceal their own expectations of reciprocation.

Bourdieu[32] elaborates upon Mauss' ideas of obligatory exchange in gift-giving, noting that this obligation exists in virtually all social relationships. He argues that social exchanges not only require labor to complete the actual exchange, but also depend on symbolic labor to *conceal* the obligatory nature of these exchanges. Thus, for Bourdieu, this symbolic labor of concealing is imbedded in social relationships, which he labels "sincere fictions of disinterested exchange."

Building on Bourdieu's work, race scholars Feagin, Vera, and Gordon[33] argue that sincere fictions exist not only when symbolic labor is used to conceal social obligations in reciprocal exchanges, but also when it is used to mask dominance in unequal power relations. In this way, sincere fictions can function to both maintain and conceal a dominant group's power over the subjugated by constructing it as normative and legitimate. These scholars note that sincere fictions can be understood similarly to Gramsci's concept of hegemony[34] (i.e., the ways in which ideologies maintain domination through willing consent) in describing how sincere fictions concurrently mask and reinforce racial inequality.

For issues of race, symbolic labor is necessary to conceal both the historical reality of the United States as a society fundamentally based upon racial oppression, and the present realities of systemic racism that maintain the current racist social order.[35] The dominant racial group constructs sincere fictions to minimize or distort the United States' bloody racist history, as well as to mask or justify the perpetuation of racial oppression in contemporary society. Sincere fictions of whiteness are thus invented notions of what it means to be white that are constructed both to assuage white guilt from slavery and other racial atrocities, and to maintain and conceal the system of racial oppression that exists in contemporary U.S. society.

Whiteness is a relatively new and understudied concept in race scholarship because it has historically been viewed simply as "normative," and an awareness of whiteness was only seen in contrast to "others." Whiteness is also relatively understudied in the media literature. However, the general perspective is that, just as stereotypes of minorities are reproduced and reinforced by media representations, so too are the norms of "goodness" associated with whiteness.[36] It is through this normalization that whiteness is not recognized as a dominant racial category but is seen instead as colorless and neutral. The normative power of whiteness rests in its invisibility and in its appearance of universality, and sincere fictions of whiteness involve symbolic labor to celebrate this perspective and to conceal its dominance.

MMORPGS AS SINCERE FICTIONS, VIRTUAL EUGENICS

The symbolic labor at work in fantasy MMORPGs, which by and large exclusively constructs white humans as heroes, presents a sincere fiction of whiteness that parallels the same practices in other media. In their work on sincere fictions of whiteness in Hollywood films, Vera and Gordon[37] note that with regard to race, films serve two dialectical functions—first, as a means to define, celebrate, and privilege whiteness, and second, as a device to conceal, obscure, and deny the existence of racialized social systems in U.S. society that produce those privileges and perpetuate racial oppression. They argue that these two functions are key in informing our understanding of how films, as commodities for consumption, produce and reproduce racism. This work is valuable not just in examining films, but in evaluating other types of media, and can contribute significantly to a greater understanding by focusing on how whiteness is constructed, how it is seen in contrast to others, and how it renders itself normative and invisible.

The symbolic significance of these acts is connected to the lengthy history of eugenics in Western society. Based upon beliefs in the cultural and intellectual superiority of whites, scholars as early as Charles Darwin[38] and his cousin Sir Frances Galton[39] argued that selective racial breeding was necessary to ensure the health and purity of the (white) human race. This is directly reflected in Said's concept of "orientalism," defined as the stereotypical, negative, and "otherized" way in which the "East" is characterized in the West.[40] It is no accident that these arguments coincided with Western colonialism, and the relationship between the two is inextricable. Representations in images and art from the colonial era contained the same themes that are associated with mainstream depictions of race across the media landscape today. These images usually portrayed a white person, most often male, in some dominant position (colonizer, teacher, etc.), and depicted non-whites in the images, regardless of their actual race, ethnicity, or phenotype, as dark-skinned savages with ape-like features.[41] It is no great surprise then that even contemporary media that borrows so heavily from symbols of colonialism would carry with it the related principles of eugenics.

How the legacy of eugenics is carried on through the construction of sincere fictions—i.e., in the propping up of the white dominant majority and the degradation and exclusion of people of color—is in no way confined to western *fantasy* media, but is widespread in all mass media in the contemporary United States. Similar representations have been, and remain, commonplace in other contemporary forms of media. Wilson et al. note that during the rise of popular media in the 1950s, traits of non-white characters were pulled directly from ages-old negative stereotypes of ethnic minorities. In general, these characters were portrayed as either the "slave" or "mammy" who was content in his/her docile position (as the case with Aunt Jemima) or the hyper-active, criminal buffoon who was good for a laugh (such as with Frito Bandito). Negative images of people of color, in contrast to positive images of whites, continue to be produced and reproduced in contemporary media. For example, media representations of black women today range from the dehumanized and dependant "welfare queen" of the 1980s to the nameless and faceless hyper-sexualized women dancing in the background of contemporary popular music videos.[42] By contrast, almost all positive or "neutral" non-white characters in popular media are white-like in phenotype (light skin, straight hair, and other Caucasian features)[43] which, as others have argued, are used as a systemic measure of beauty within non-white communities.[44] In this way, whiteness is contrasted with blackness and is assigned moral, intellectual, and physical superiority.

In fantasy MMORPGs, white protagonists engage in heroic battles and emerge victorious over evil. Within these narratives, they are depicted as the embodiment of bravery, strength, and moral character; their completion of herculean feats earns them well-deserved glory and esteem. By contrast, non-white characters exist in heavily stereotyped and adversarial ways, or they are rendered largely nonexistent within these virtual worlds. In this way, whiteness is indeed "propped up," while racial oppression is symbolically reflected and reinforced. MMORPGs thus provide little to no emancipation from the systems of racism that exist in society at large, but rather maintain the dominant racial ideology in ways similar to other media forms. Unless the designers of MMORPGs develop a more critical perspective in how the contemporary racial order insidiously penetrates these virtual worlds, the hope for MMORPGs to free individuals from constraints of society at large will never be realized.

CONCLUSION

It is our intent in this chapter to advance the dialogue on MMORPGs and social inequality, particularly in the ways that many of these games interface with racial oppression. These virtual worlds, like so many other forms of media, possess a tendency to privilege whiteness while concurrently rendering that privilege normative, and (as a consequence) invisible. This likely carries profound implications for our understanding of the emancipatory nature of these games. Players possess differential levels of agency within these virtual worlds based upon their racial category in the "outside world." Even if virtual worlds allow opportunities for people of color who play fantasy MMORPGs to in some sense "transcend" their race, this transcendence is contingent upon their taking on a white identity and successfully "passing" as white.

While the focus of this chapter is primarily on how these games are "set up," readers will inevitably raise questions about how their racial elements are actually interpreted by gamers. It is important to emphasize that we do not argue that fantasy MMORPG players or readers of Tolkien and C.S. Lewis generally identify the colonial and racial allegory as racist and embrace it as such, and we do not argue that fans of the high fantasy genre consciously and explicitly celebrate its racial elements. In fact, we propose quite the opposite. As we have noted, one primary function of white privilege is to shield members of the dominant group from seeing how dominance is manifested in social worlds (virtual or not). Consequently, members of the dominant group are rather unlikely to possess an awareness of how racial dominance permeates these forms and, like the players in the previous examples, are much more likely to see them as "raceless" places. Thus, the racial components endemic in the high fantasy genre—including fantasy MMORPGs—function, intentionally or not, to maintain the dominant perspective of race while appearing color-blind, benign, and normative.

Having said this, future research in this area could indeed contribute to the ongoing dialogue on race and MMORPGs by focusing on the extent to which media consumers negotiate the racial representations outlined here. This could be accomplished through a number of different social research methods, including participant observation, focus groups, and in-depth interviews. As scholars[45] have suggested, employing qualitative methods can successfully reveal the nuances of "race talk," or the ways in which people discuss issues of race and the extent to which their perceptions reflect dominant racial ideologies. In fact, it is these dominant ideologies that provide the basis for perpetuating white privilege in the United States and greater Western society. Future research could thus provide important insights into how these ideologies are maintained in MMORPGs, as well as how they can be resisted and restructured in ways that truly pursue emancipatory interests.

NOTES

1. For just a few examples, see S.G. Jones, *Virtual Culture: Identity and Communication in Cybersociety* (New York: Sage, 1998); D. Williams, "Why Game Studies Now? Gamers Don't Bowl Alone," *Games and Culture 1*, 1(2006): 13–16; N. Yee, "The Demographics, Motivations and Derived Experiences of Users of Massively-Multiuser Online Graphical Environments," *Presence: Teleoperators and Virtual Environments 15* (2006): 309–329.

2. D. Williams, N. Ducheneaut, L. Xiong, Y. Zhang, and N. Yee, "From Tree House to Barracks: The Social Life of Guilds in World of Warcraft," *Games and Culture 1*, 4(2006): 338–361.

3. N. Ducheneaut, N. Yee, E. Nickell, and R. J. Moore, "Alone Together? Exploring the Social Dynamics of Massively Multiplayer Games" (paper presented at the annual meeting of CHI, Montreal, Canada, April, 2006).

4. D. Thomas and J. S. Brown, "Why Virtual Worlds Can Matter," *International Journal of Learning and Media 11*, (2009): 37–49.

5. M. Wertheim, "The Pearly Gates of Cyberspace." In *Architecture of Fear*, ed. by N. Ellin (New York: Princeton Architectural Press, 1997), 295–302.

6. Neal Stephenson, *Snow Crash* (New York: Bantam Dell, 1993).

7. J.R. Feagin and H. Vera, *White Racism* (New York: Routledge, 1995). See also H. Vera, J.R. Feagin, and A. Gordon, "Superior Intellect? Sincere Fictions of the White Self," *Journal of Negro Education 64*, 3(1995): 295–306.

8. E. Goffman, *The Presentation of Self in Everyday Life* (New York: Anchor Books, 1959).

9. C. Cheung, "Identity Construction and Self-Presentation on Personal Homepages." In *The Production of Reality: Essays and Readings on Social Interaction, Fourth edition*, ed. by J. O'Brian (Thousand Oaks, CA: Pine Forge Press, 2004), 310–320.

10. For two of the original arguments on this, see H. Rheingold, *The Virtual Community: Homesteading on the Electronic Frontier, Second edition*(Cambridge, MA: MIT Press, 2000), and P.C. Adams, "Cyberspace and Virtual Places," *Geographical Review 87,* 2(1997): 155–171.

11. S.G. Jones, *Virtual Culture: Identity and Communication in Cybersociety* (New York: Sage, 1998).

12. T. Higgin, "Blackless Fantasy: The Disappearance of Race in Massively Multiplayer Online Role-Playing Games," *Games and Culture 4*, 1(2009): 3–26.

13. Higgin, "Blackless Fantasy," 8.

14. G. Lastowka, "Cultural Borrowing in WoW," (terranova.blogs.com, May 16, 2006), accessed at http://terranova.blogs.com/terra_nova/2006/05/cultural_borrow.html.

15. For authoritative works on race and racism in virtual worlds, see Nakamura, *Cybertypes* (New York: Routledge, 2002). See also Higgin, "Blackless Fantasy."

16. See, for example, S. Kim, "Beyond Black and White: Race and Postmodernism in *The Lord of the Rings* Films," *Modern Fiction Studies 50*, 4(2004): 895–907.

17. See Kocher's extensive review of Tolkien's works in *Master of Middle Earth: The Achievement of J.R.R. Tolkien* (London: Thames and Hudson, 1972).

18. Thomas Wagner presents a detailed critique of the anti-Arab and anti-Muslim sentiments in Lewis' final book of *Chronicles of Narnia, The Last Battle*. See T. Wagner, "The Last Battle" (sfreviews.net, 2005), accessed at http://www.sfreviews.net/narnia07.html.

19. Kim, "Beyond Black and White."

20. M. Omi and H. Winant, *Racial Formation in the United States: From the 1960s to the 1990s* (New York: Routledge, 1994).

21. Higgin, "Blackless Fantasy," 11.

22. This information can be found on the *World of Warcraft* website, accessed at www.worldofwarcraft.com.

23. *World of Warcraft*, www.worldofwarcraft.com.

24. *World of Warcraft*, www.worldofwarcraft.com.

25. *World of Warcraft*, www.worldofwarcraft.com, emphasis added.

26. *World of Warcraft*, www.worldofwarcraft.com.

27. *World of Warcraft*, www.worldofwarcraft.com.

28. *World of Warcraft*, www.worldofwarcraft.com.

29. Castranova's (2005) blog post entitled "The Horde is Evil" and Lasowka's (2006) post entitled "Cultural Borrowing in WoW" explore issues of good and evil in *World of Warcraft* and address many of the characteristics of in-game races described in this paper. These two posts sparked a debate (containing over 1,000 responses) over the ambiguities and complexities of the *World of Warcraft* storyline, with some individuals arguing that the game's creators may be constructing a narrative that symbolically exposes the evils of Western colonialism.

30. Higgin, "Blackless Fantasy."

31. M. Mauss, *The Gift* (London: Cohen & West, 1966).

32. P. Bourdieu, *Outline of a Theory of Practice* (Reading, MA: Addison-Wesley, 1977).

33. Feagin and Vera, *White Racism*; Vera, Feagin, and Gordon, "Superior Intellect?"

34. A. Gramsci, *Selections from Prison Notebooks* (London: Lawrence & Wishart, 1971).

35. For a concise analysis of the United States' history of racial oppression and the symbolic labor required to conceal racism, see J. Feagin, *The White Racial Frame: Centuries of Racial Framing and Counter-Framing* (New York: Routledge, 2010).

36. R. Dyer, "White." *Screen 29*, 4(1988): 44–65.

37. H. Vera and A. Gordon. *Screen Saviors: Hollywood Fictions of Whiteness* (Lanham, MD: Rowman and Littlefield, 2003).

38. C. Darwin, *On the Origin of Species by Means of Natural Selection* (London: W. Clowes and Sons, 1859); C. Darwin, *The Descent of Man* (New York: American Rome Library Company, 1902).

39. Sir F. Galton, *Inquiries into Human Faculty and its Development* (London: Macmillan, 1883).

40. E. Said, *Orientalism* (New York: Pantheon, 1978).

41. See for example, A. McClintock, *Imperial Leather: Race, Gender and Sexuality in the Colonial Contest* (New York: Routledge, 1995), and C. Wilson et al., *Racism, Sexism and the Media: The Rise of Class Communication in Multicultural America, Third ed.* (Thousand Oaks, CA: Sage, 2003).

42. M. B. Littlefield, "The Media as a System of Racialization: Exploring Images of African American Women and the New Racism," *American Behavioral Scientist 51*, 5(2008): 675–685.

43. A. Cortese, *Provocateur: Images of Women and Minorities in Advertising* (Lanham, MD: Rowman and Littlefield, 2008).

44. P.H. Collins, *Black Feminist Thought: Knowledge, Consciousness and the Politics of Empowerment, Second ed.* (New York: Routledge, 2000).

45. K. Myers, *Racetalk: Racism Hiding in Plain Sight* (Lanham, MD: Rowman and Littlefield, 2005).

Chapter Nine

The Goddess Paradox

Hyper-resonance Shaping Gender Experiences in MMORPGs

Zek Cypress Valkyrie

When I began playing *Final Fantasy XI*, I created two avatars, both "Hume," serving the nation of Bastok, but one was a female avatar and one was a male avatar. As I foraged through the virtual world casually switching between avatars, attempting to level them equally, I took notice of how the players treated me. As "Reimi," finding parties was easier, getting help with my sub-job quest was simple (completed in one night), and a Tarutaru Black Mage I met in Valkrum Dunes even gave me 5,000 gil, which I was not expected to pay back. As "Edge," the game was more difficult. Parties were harder to come by, and less likely to last. The subjob quest took days to complete, and no one offered to hand me free gil. [1]

The classic role playing game (RPG), *Lunar: Silver Star Story*, [2] weaves the tale of the reincarnation of goddess Althena into the talented, beautiful songstress Luna and the quest of dragonmaster hopeful Alex's journey to save her. Goddesses in the fantasy genre have been both idolized and infantilized. Despite Luna's heritage and seemingly endless power, she is merely a fragile girl who is kidnapped by the Magic Emperor and used as a sexualized pawn. Only Alex's heroic quest and his assumedly chaste love will save Luna and allow her to simply live life as a normal girl. Similarly, women who play Massively Multiplayer Online Role Playing Games (MMORPGs) [3] are confronted with a paradoxical gender experience. Research on gender has often highlighted paradoxical attributes to how people experience their gender (i.e., men feeling powerless as individuals though they are powerful as social category). [4] Although the solid world, [5] that is, the world outside the game may seem to diminish in importance within the virtual world, players ardently carryover their solid world statuses into virtual spaces. As with other realms that marshal an almost prophetic bluster to rewrite how gender *could* be done, gaming worlds are similarly problematic. MMORPGs and the players within them continue to (re)produce the rigid gender dichotomy despite the potential for new layers of flexibility. In doing so, women who participate in these virtual spaces are treated as the male-dominated space dictates. Women are welcome exceptions, treated warmly, assisted openly, and befriended readily. And yet, seen more as potential romantic and sexual partners than as players, they are harassed, infantilized, and never fully embraced as gamers.

It is important to note that MMORPGs, and the avatars[6] within them, are not created in a vacuum. Taylor[7] and Higgin[8] have argued that the limitations of avatar options in respect to gender and race may reinforce culturally idealized bodies and curtail players' ability to choose what they want in virtual worlds. This is evident in the aesthetic appeal of the gendered avatars: females have thin waists, exceptional breasts, and curvy hips, while males have rippling abs, swollen biceps, and chiseled features. Implicit in this argument is that game developers operating on these conventions limit players to only certain choices with regard to avatar appearances, limiting players from creating their *own gendered* character appearance. Game designers are also constrained by technological and economic rhetoric that encourages them to market to very particular audiences and thus aim for "bestseller" titles.[9] This means that MMORPG worlds and avatars are built in the solid world, and though they exist in the virtual, they are hardly created using a "blank slate." This contrasts somewhat with the predecessors of MMORPGs, Multi-User Dimensions (MUDs), which arguably allowed for more creativity on the players part in "avatar" creation. Specific to gender, players of MUDs could select gender "neutral" characters or "morph" between genders.[10] Although MUD players could question the meaning of such selections, no such neutral or undefined area exists in MMORPGs where avatars are hyper-gendered and players merely select one of two options.[11]

Taylor argues that MMORPGs have created deep and meaningful experiences for women gamers. MMORPGs, although constrained by the "gamer" (read: male) stereotype and market demographics,[12] appeal to women players for a multitude of reasons. According to Taylor, women gamers are drawn to aspects of power and status, identity play, multifaceted social immersion, and game mastery. These experiences can be inhibited by avatar design, which reinforces idealized (and sexualized) female bodies. Although women gamers experience various degrees of virtual success, the gender identities they perform and their interactions within virtual worlds can alter their experiences of and their access to that success. Negotiating gender in these virtual spaces is a process embedded in a history which wove gender and technology into a predominately male tapestry.[13] Because gaming continues to be presented as a male space, the existence of "real women" in such spaces is clearly labeled as exceptional. Players (men and women) seek to establish those claiming to be women as "female-bodied." They believe both that they can assume body from behavior[14] (a gendered body behaves in respect to said body, thus through recognizing behavior one can assume the correlating body) and that gender ("doing" femininity)[15] is always congruent with sex (socially agreed upon criteria which categorize a person as female). Women are thus challenged to reinforce these assumptions in game worlds, establishing themselves as women by marshaling evidence of their solid world gender (for example, voice chat with other players or providing photographs as verification). In contrast, men need not prove to be male-bodied. Embodied gender theory[16] argues that the norms for striking balance between sex appearance and gendered behavior vary in different contexts and are enforced by those with power in those contexts, but that men and women's bodies often carry different meanings or are given different weight in the same context. This male-arena treats men's (solid world) bodies as given (though invisible) so long as men's gendered behavior is normative and appropriate. Women's (solid world) bodies, however, are imperative to establishing the authenticity of their gendered behavior, and furthermore their gender identity (again, the body is not always a requirement but a correlate of what players assume to equal a female body such as a "feminine voice" heard through voice chat in game allows players to assume the gender-sex link).

I argue that game worlds marshal an "approach" to playing the game that demands players be transparent about their solid world statuses. In doing so, the virtual world loses the potential for gender flexibility and, in turn, attempts to reinforce gender assumptions within this ethereal realm. Players explain that gender is paramount to how a person is treated within the game, and reason that the process of identifying "real women" is in effect to be able to treat the respective genders as they should be treated in the virtual world. The consequences for women who "prove" to be female-bodied ironically grants them sexual and romantic power over men, while allowing (if not encouraging) men to infantilize women within the game as inferior players, thus the goddess paradox.[17] This concept not only requires that women's (proclaimed) bodies be the conduit of both their empowerment and disempowerment, but that both come as a consequence of a form of masculinity that mandates "heterosexual desire."[18] This adheres to tenants of social exchange theory which hold that women gamers become invested in proving their gender to access rewards from men while simultaneously reinforcing more traditional gender roles and identities of both the men and women.

METHODS

My interest in player treatment with respect to gender began with my own immersion into MMORPG worlds. *Final Fantasy XI Online* (*FFXI*) released in 2003 in the United States, was the first virtual world in which I became heavily invested. I spent five years in *FFXI* and continued to play a variety of other MMORPGs including, but not limited to, *World of Warcraft* (*WoW*), *Final Fantasy XIV* (*FFXIV*), *Flyff*, *Rose*, *Sword of the New World*, *Lineage 2*, *Requiem*, *Warhammer*, *Ether Saga*, *Shin Megami Tensei*, *Florensia*, *Ever-Quest* (*EQ*), and *Guild Wars* (*GW*). I also worked as a salesclerk for GameStop from 2000 to 2006. During this time, I spoke with dozens of MMORPG players about their online gaming experiences as they purchased software. In addition, I have been an avid RPG fan for the last fifteen years (what most consider the predecessor genre from which MMORPGs emerged). Through this foundation, I was able to establish solid connections with MMORPG players as I accompanied them in their virtual exploits.

I have logged approximately 500 days[19] of gameplay in the aforementioned titles over a five-year period, although I draw heavily from my immersive experiences in *FFXI* (250 days) and *WoW* (150 days) because I spent far more time in "Vana'diel" and "Azeroth" than I did in other worlds. I also felt a sense of "game mastery" in both of these games, which was evident by my participation in "end-game" activities and my deep knowledge of the areas, quests, classes, crafts, monsters, and so on of both worlds. Game mastery, I felt, was vital in understanding how players felt and related to elements of the game. In other words, it gave me a good sense of what these games were about and the complex relationships within them as they related to game design. Game mastery also made interviews and conversations with players fruitful because I could easily compare my own experiences with theirs which gave me a deeper understanding of the data. I cannot sufficiently stress the importance of spending a large amount of time in this setting.[20] I nurtured my relationships with other players through extensive quality time in-game. I befriended players, spent time chatting informally, and assisted them in their gaming goals (leveling, completing quests, and farming). My extensive "insider knowledge"[21] of games and gaming allowed me to foster these relationships.

I combined my observations, field notes, informal interviews, and experiences in situ with semi-structured interviews outside the setting. I conducted interviews with 50 MMORPG players (14 female players and 36 male players); interviews ranged from one hour to four hours each. The age of participants ranged from 18 to 41, and the majority reported logging at least 100 days in one or a combination of MMORPGs. Most interviewees lived in the United States (41), although some were from Australia (2), Mexico (1),

Canada (2), and parts of Europe (4). I recruited participants predominately through the relationships I built with them in the game although I recruited a handful of interviews (8) through my connections with GameStop and on MMORPG message boards. Interviews were conducted over several mediums including face-to-face (14), telephone (7), and instant message [22] (29) programs. Kvale [23] argues the interview process should remain reflexive and dynamic, promoting the interaction with each participant while guiding the interview thematically toward the root of the investigation. I utilized a themed, semi-structured interview guide with a built-in flexibility that reflected the experiences of each participant. The venue of the interview was their choice as I wanted to ensure their comfort. Telephone and face-to-face interviews were recorded with a digital recorder and later transcribed and deleted. IM (instant message) interviews were duplicated into a separate file and transcribed, eliminating screen names.

I coded my data into broad topics of interest through a process of reading and re-reading the transcripts and organizing data into useful categories. [24] I continued coding and analysis as I was collecting data, allowing me to augment my interview guide and solidify my evolving conceptual models. Transcribed interviews and field notes (as well as transcribed screenshots) were pooled and coded for themes of interest in a process of "data reduction." [25] I organized my data into "talk" about topics that might help answer my research questions. Using general headings in multiple "Word" files, I created aggregate files for whole themes and categories. These umbrella themes were initially coded for ideas and phrases of significance which became more focused as I continued to collect interviews and spent more time in the setting. Interviewee responses allowed me to narrow my focus on each theme. These themes were further organized into sub-concepts that allowed me to ask more specific questions in later interviews: a process of reflexively analyzing the data and continuing to augment the thrust of my questions. Eventually, I experienced diminished novelty in data collection, which arguably suggested that I had reached theoretical saturation and had a more complete set of codes for my models.

REWARDING SOLID WOMEN

Players who managed to "prove" to be women (that is, they were believed to be and accepted as women) were confronted with significantly different gaming experiences than those assumed to be men. Although the process for "proving" gender itself could be discussed in-depth, the following discussion centers on the experiences and perceptions of those players who were accepted as "real" women. Part of the reality in this male dominated space was that "real woman gamers" were welcome "exceptions" to the virtual world and were treated as such. Women gamers were afforded benefits which included in-game assistance, money, parties, friends, attention and conversation, and "nicer" treatment. However, they had to negotiate a number of detriments which included unwanted romantic and sexual advances, perceptions of lackluster gaming ability, and encapsulated roles.

PLAYING IN HYPER-RESONANCE

Paradigms of play among gamers are as varied as the players themselves. However, players largely fall into one of far fewer "approaches" to what can be considered *world resonance.* [26] World resonance is quite simply the "slippage" between the "solid" (often misleadingly dubbed the "real") world and the virtual (also called the synthetic or game) world. World resonance can determine how a player negotiates and controls information about their solid world statuses within the virtual world. For example, players with "greater overlap" or higher resonance between worlds as it were (hyper-resonance, or hypers for short), [27] maintain far more consistency in their solid-to-virtual world identities. This is the "be yourself" approach that has become an almost inarguable standard in online worlds as they continue to grow into the mainstream. Hypers believe in authenticity and consistency, and frown upon experimentation. Hypers embrace a more rigid form of play online, but marshal it is the more "honest" form. This standard was experienced by Vincent, [28] a 23-year-old SM/VM [29] (solid man/virtual man):

> At first, I thought I was supposed to role-play when I played. I was trying to act like a warrior character would act. Protect the mages, be fair, and be chivalrous. But when I realized everyone else was just being normal, the roleplay went bye-bye and I just started talking and acting like me.

Similarly, Freya, a 22-year-old SW/VW (solid woman/virtual woman) explained:

> I like to be just myself so my character does and acts exactly how I would act normally, minus the spellcasting, throwing shurikens at stuff bigger than me, rising from the dead after getting beaten up and drinking poison potions for fun in Jeuno . . . the provoke thing I can do however. I do not understand well people who would like to role-play. . . . I think this should apply to a lot of people, playing a video game or not, you are dealing with real people.

Although role-playing is part-and-parcel of the MMORPG's experience, many players adhere to the hyper-resonance standard, that is, one should be who they "really" are (exceptions are the obvious fantasy context: spellcasting and monster slaying is permissible). For the sake of this argument, it is important to establish hyper-resonance as the preferred style of play as it directly relates to gender interaction within the game. When players openly acknowledge a hyper-resonance standard, gender play is cast out. The solid world is in effect transposed upon the virtual with little disturbance. Those adhering to a different approach (for instance RPers)[30] are less concerned with solid-to-virtual consistency (albeit this is contestable) and therefore are not at the center of this discussion.

Hypers enjoy creating avatars as "extensions" of themselves as opposed to "new personae" as Penelo, a 22-year-old, SW/VW wrote:

> Always my same gender. Well I guess when I make a character (when I made Penelo), I want her to look as much like me as possible in terms of both physical and feelings. So I picked a Tarutaru because I felt like in real life I was cute and cuddly so I made her with brown hair and as short as possible to resemble me physically, an extension of myself in game. I don't think I could identify with a guy character who looked nothing like me. I like being my character not just HAVING a character.

Again, it is important to reiterate that hypers acknowledge the presence of difference between themselves and their avatars (they know, for example, that they are not "elves"), but remain consistent in presentation of themselves (their avatar is more of a "funnel" for their solid world identities and not a "barrier" preventing others from knowing their solid world selves). Hypers in no way attempt to take on a new gender persona or identity in the virtual world and, possibly just as important, they hold other players to this same standard. As noted earlier, this standard has become increasingly rooted in the MMORPG experience as these games have attracted more and more players. Ivy, a 22-year-old, SW/VW explained:

> I think people's interactions on *WoW* are just a microcosm of real-life interactions because the game is more mainstream than others. I think in other MMORPGs, it might be easier to adopt the role-play attitude and treat characters based on the character's gender rather than the person's gender because they appeal more to a subset of people who are dedicated to the role-play thing. And keeping in-character is one of the things role-play is about. Since *WoW* has such a huge variety of people playing it, I think it's going to be more like any social situation where your actual gender can determine how you interact with someone.

Ivy's sentiment highlights several aspects of the modern MMORPG experience. First, that approaches to play possibly vary by game (or by server). However, she also explains that the non-hypers are a "subset" of people. In other words, the hyper approach, that is the non-RP, "just like real life," "authentic extension" approach, is the preferred and more popular approach to gaming online. Because of this, solid world gender is required to be as transparent as possible. Players openly acknowledge difference in treatment by gender, and marshal that the difference in treatment is part of the reason why players *should* ascribe to the hyper approach which demands they disclose, or better yet, remain consistent in gender-of-avatar choice as it should be congruent with their solid world gender. This, in turn, limits players from broadening the gender spectrum. Because the rigid gender dichotomy is readily enforced and is coupled with an authenticity norm those desiring to embody alternative gender identities are largely relegated to "exceptional spaces" in virtual worlds such as servers or guilds dedicated to (or, at least accepting of) RP and gender play.

REAPING BENEFITS

Superficially, it seemed that women gamers experienced a number of benefits when they chose to participate *as women* in the virtual world. This meant that these women were seen as women in the eyes of other players (and may have been required to prove they are indeed women). Once the status was in place, "hyper-women" (women who had revealed their gender and were also player-to-avatar gender consistent) were then coveted for their "rarity" when compared to "overflow" of men in gaming worlds.

TREATING WOMEN AS WOMEN

Both men and women reported that players were treated differently in game on the basis of gender. Laike, a 31-year-old SM/VW *WoW* player, wrote:

> For whatever reasons or motivations, I think a lot of males treat females preferentially. I usually see this in guild settings, where people see or talk to each other on a frequent basis. Sometimes there will be a particular person that won't be denied any requests for help, or will always be provided with money or items as desired.

Gin, a 24-year-old SM/VM resonated:

> Without a doubt. If a female player lets on that she is, she'll tend to get more attention. I know of some female players that make male characters to avoid that. Other players may come to them more readily to help them or will come on to them more often.

Hyper-women were treated as exceptional in the virtual world. Gaming culture has continued to be cited as a male-dominated pastime. However, the accuracy of this claim (whether more men than women played these games) was not as important as how the perception of that reality (believing women were rare) framed the experiences of women who chose to participate in these games. Perceived as exceptions, hyper-women accessed a novelty status that granted them "special" treatment. Since the perception of this treatment superficially seems to provide a number of benefits to women gamers, hyper-women became invested in ensuring they could access their novelty status. Because of this, women were invested in proving they were women to increase their chances of in-game success. Penelo wrote:

> To be honest I think I was way better off being female in people's eyes because men are taught to always help women and be chivalrous. That was sooooo apparent in game. I mean as long as I was nice to male characters in game, if I said I *really* needed help or I *really* had no money, they would take me places and help me or buy me things, sounds devious, but it was *awesome.* Meanwhile, my boyfriend was always having to beg people for help, so I felt lucky but it's not like it was exploitive to me, I was fine with having a pic up (posted online) and people being super nice to me and the guy players got to feel helpful win-win right? Hahahaha.

Women playing virtual games were empowered over men with (if nothing else) the ability to obtain more help in-game in the form of other players (mostly men) providing gifts (money, equipment) or assisting them with their virtual exploits (killing monsters, experience parties, questing, and so on). Penelo compared her experience playing the game with her boyfriend's experience highlighting that he was far less likely to obtain the assistance required to complete some elements of the game (or that it took considerable effort on his part to receive the same level of help as Penelo received). She continued:

> Well for example, for my white mage spells I had to do these really hard quests or pay a crap load of money to get them, but once I complained enough (not annoying . . . just disappointed) people bought them for me [laughs] . . . whereas I remember my bf working his ass off to get things like AF done and get certain quests done. So yeah, it was advantageous to be a female in my opinion (as long as you were REALLY female) cause if you said it and couldn't prove it you were as good as dog chow.

Penelo cited two realities of in-game play that were important to accessing this "preferential" treatment. First, she, as well as other women, was required to prove she was a woman. Second, she almost consciously performed a "damsel in distress" or submissive femininity in order to solicit assistance. In contrast, men in game worlds are expected to be proficient, if not masters, because the stereotype of the (male) gamer disallows men from receiving the same amount of help (although temporarily, many gamers can claim a "noob" status that, while stigmatized, can be deployed as an adequate excuse for lack of game mastery). This did not disqualify women from more active and even powerful roles, but presented the performance and adherence to a submissive femininity as a path of least resistance.[31]

Although it is arguable whether this "better" treatment is completely positive, in MMORPGs, where progress and status in the game are unavoidably linked to forming relationships and cooperative play, being a hyper-woman was advantageous. Being able to obtain more assistance in-game allowed hyper-women to excel in meeting their in-game goals (completing missions, felling infamous monsters, obtaining valuable equipment). This meant that hyper-women were encouraged to perform a submissive femininity in the virtual world, which allowed men access to forms of masculinity that embodied a brand of "virtual chivalry." This symbolic reconstruction of an eroticized "heterosexual desire" reified already problematic gender power differences in these new virtual spaces.

STREAMLINING GAME MASTERY

After extensive time in-game, most players cited their virtual accomplishments as evidence of their game mastery (possessing certain items or equipment, participation in difficult battles, completion of particular quests, titles, level of their crafts, classes, or skills). Because the perception among players was that hyper-women were helped more often, some men lambasted the women's approach to playing the game as did Vincent:

> I think females get helped more. I had this guildie that was playing a female character who was "dating" a male character in the guild. She basically used him and made him help her by playing the "oh noes I can't do it myself" act. He helped her get drops, claim NMs, do the missions, do her AF quests, EVERYTHING (yells). I literally had to beg for assistance with anything I couldn't solo, which in *FFXI* is almost everything. I think female characters have it easy because they are seen as potential girlfriends to all the gamers out there.

Feeling threatened, some men (and women) lash out at hyper-women for "using" male players to meet their in-game goals. Many players like Ryuk, a 21-year-old SM/VM *FFXI* player, commented that men who "couldn't get any in real life" were seen as "coming to the aid of" women who could possibly become romantic partners. These "benefits" allowed women to make faster and easier progress in terms of game mastery. Interestingly, men who perceived hyper-women as receiving more assistance attacked the men who were providing the assistance and largely ignored that part of the problem lay in men *not being helped* and *not helping* other men.

Coincidentally, some hyper-women who witnessed the same disparity in treatment and assistance instead blamed the women who were accepting the preferential treatment. Freya, a *FFXI* veteran, protested:

> I do believe that men and women are treated differently in-game. Usually mentioning the fact you are female will give you a huge advantage with guys looking for a bit of E-pussy. For the record: I have not and never will exploit this, it's probably one of the most degrading and sickening things a woman can do and makes her no better than a street corner whore. . . . A lot of the women in game do exploit this for items, gil, etc., and it sickens me that a person would lower themselves to such levels for pixels. . . . I would say it's an unfair advantage over men, since some men are so besotted with "women" they meet in game that they'd get them through harder missions, pretty much play the game for them if it got them somewhere . . . which is sad since it defeats the purpose of the game and dulls a lot of the finer points.

Some women players argued that "besotted," "love-starved," or "desperate" men were in essence ruining the game by "padding the grass" for women gamers to reach all their goals (all with intentions to pursue these women romantically or sexually, as if to say men will not help women without compensation). Subsequently, women who were seen as "exploiting" the men were subject to criticism by other men (who struggled to obtain the same kind of support) and women (who accused them of "whoring for pixels").

Implicit in the reality of women's pseudo-empowerment over these "geek men," was that it also threatened relationships in the solid world. Some men cited feeling anxious about their partners (or other men's partners) using their "feminine appeal" to enchant the "uber dorks." Magnus, a 25-year-old, *WoW* player explained:

> That happened to a guy I know. He said he came out in the living room one night and his wife was playing, but webcaming with another guy via the PS Eye thing. I was like yep, that's why chicks shouldn't play games. Like I actually felt for that guy. . . . It's not the playing without me thing that I care about. It's what's being SAID while playing. I could give a damn if anyone [speaking of his partner] is playing a game without me. As long as it's fucking appropriate. What I don't understand honestly, is the fucking uber dorks that go after these women. . . . With the enormous amount of porn on the internet, and getting laid now as easy as walking outside, it doesn't make sense to me.

Though many men (and some women) couched their criticism of hyper-women's augmented game experience due to the attention lavished upon them as being "unfair" or as providing them with an "easy game mode," few acknowledged underlying threats to both claims of masculinity and femininity. In other words, men feared other men's virtual accomplishments as providing them an advantage to pursuing hyper-women (or worse, "stealing them" from their solid world partners) while women feared being forced to juggle their identities as gamers with negotiating an overzealous male majority that would not see them as anything but potential partners.

NEGOTIATING DETRIMENTS

Despite being catered to by men in-game, hyper-women were often subject to continuous romantic and sexual advances. Additionally, having accepted assistance from men (or having to carry that perception) hyper-women's gaming prowess was often questioned.

Sidestepping Sexual and Romantic Advances

The constant and often invisible maintenance of unwanted advances by hyper-women was most profoundly stated by Xenobia, a 27-year-old SW/VW who wrote to me in-game about her problems with her guild members:

> So I have a 12-year-old, a 14-year-old, two 16-year-olds, and a 22-year-old who claim to be in love with me, cuz I'm "so amazing and beautiful." Do you know how much of a headache it is to try and not hurt their feelings? What the hell am I doing wrong? Why can't I play like anybody else without people falling in "love" with me? One kid told me he masturbated to the thought of me. He's the 12-year-old. I felt like a CHOMO [child molester]. I told them I'm too old to be anything but a friend to them and I have a man in my life so it's not gonna happen, erg, what do I do? I don't want to be mean about it, but it kind of creeps me out.

Women had to balance the perception that they were always seen as possible romantic interests, with their desires to meet gaming goals. Xenobia struggled with assisting and being assisted by guild members who lusted after her. Some women noted the sexual harassment in-game had changed their approach to playing and who they played with, remaining ever conscious of the potential consequences of engaging in romantic and sexual relationships online.[32] Near, a 23-year-old *WoW* player commented "I think guys can be animals towards the fairer sex. There was this one guy, he did an emote that said 'char name grabs char name's boob.' And the girl char, or so I would assume, did /no and /slap, just things like that." To be "like anybody else" and to circumvent such harassment was difficult for women in the virtual world and inhibited them from being considered simply gamers or "allies in play." Such romantic and sexual advances were not a "problem" for men in the games, since men did not have to justify or struggle to be perceived as *just gamers*.

Though men were often the source of such harassment, they were also conscious that such harassment took place. Men playing women (SM/VW) experienced this firsthand if they attempted to avoid hyper-resonance play. Magnus related:

> There are a lot more women these days, which is good and bad. Of course, for me the women make better friends in game cuz they are less [obsessed] about being all badass. The women in the games are usually down for helping and shit too, and are usually really cool about loot and stuff. Playing with guys, I swear, it becomes a dick measuring contest most of the time or worse. I had one guy in *WoW*, help me because I was playing my BE Female Mage . . . and then afterwards [he] asked me to cyber with him. Lol. When I told him I was a guy, he disappeared. Yea, so guys in games, a lot of them are really [idiotic]. And like it or not, gaming is still male dominated, which means [idiot] over-flow.

The problem with virtual spaces maintaining a hostile environment even for female avatars alone (as Magnus was not a hyper-woman) can repel many women (and some men) who might wish to play these games. Women, faced with relentless sexual advances may "roll" (choose) male avatars to avoid this problem, even if they would rather play as women. Men may also avoid choosing female avatars knowing that other men may eroticize their avatars and likewise make unwarranted advances. Allowing men's behavior to control players' choice of avatar or even their choice to play the game at all, obstructs the metamorphic potential that gaming, and online worlds in general, have heralded.

Lackluster Player Perception

Because assistance was (seemingly) so readily available, this often reinforced players' beliefs that women were not skilled gamers. Penelo explained, "They want to know what gender you are so they know if you 'need help' or if they can flirt with you or if they should be competing with you." Similarly, Cerberus, a 21-year-old open SM/VW on hiatus from *FFXI*, explained:

> Females got more [assistance] because of the idea males have that they're helpless and need the help, which is for the most part true, or maybe females are just lazy, but everyone has their strong points. I'm sure guys got help too, I've helped plenty, but I usually didn't ask for assistance.

This perception crystallized stereotypes about who "real gamers" were, which is to say, men. Women were welcome insofar as they were objects of desire and allowed men to execute virtual chivalry, like Beck, a 23-year-old open SM/VW *FFXI* player, who wrote:

> I know that when I'm around a female player I tone my language down a little and am a little more helpful towards them. I actually once went so far as to get a fellow mithra Fenrir (her name was Rena in game and she was from Italy). Anyway, she was a noob SMN and wanted some tips and pointers. Instead, I got her Fenrir. I pulled a lot of strings. I used many of my in-game contacts and friends and spent a lot of time doing that.

The extent of Beck's beneficence toward Rena changed her experience in-game. Fenrir (a powerful monster in the game) was a very difficult battle usually reserved as an end-game accomplishment (veteran players grouped together to defeat the monster in order to obtain it). *FFXI* was a "multi-class"[33] MMORPG, this meant one avatar could take on many different classes (Paladin, White Mage, and so on) by "switching classes" (a command in-game). Because of this, most Summoners (a class in *FFXI*) who possessed Fenrir had most likely leveled another class (for example, a Paladin) to level 75 (the level-cap at the time) and had challenged the monster with a group of powerful allies as that class. Those who had done "ao" were able to play their Summoners having possession of Fenrir, and this was rather prestigious (and, in some cases, led to more invitations to group). However, lacking a class that would have actively participated in the fight (as in Rena's situation) highlighted that someone had completed the battle *for her* rather than she had taken an active role in the fight. This reality was commonly exposed in group settings where a player like Rena would be asked what was her "main class" (i.e., with what class did she fight Fenrir) and, lacking a class that would have been able to participate in the fight, those inquiring would understand that her Fenrir was obtained through a favor such as the one Beck eagerly provided.

Some men's preferential treatment of women was done with the best of intentions. It allowed men to perform masculinities they may have had less access to in the solid world, and it contributed to the "spirit" of these games, which embodied cooperative play and altruism (although they have a number of very competitive aspects as well). It would be misleading to present all beneficent deeds towards women as sexist or lustful. However, this treatment contributed to women being pushed into encapsulated roles and thus they were not fully embraced by the gaming world. Corcell, a 27-year-old SM/VM wrote: "I tend to expect more out of male players than female players. I will curse out someone for doing something stupid if that person was a guy, while if it was a female player, I just shrug it off." The reality was that men did not expect women to even be capable of playing at a high level. Mia, a

27-year-old SW/VM resonated, "I think it's the same as out here (real life) too. Let's say you die, they think you are a chick. . . . I think it's just easier to blame girls or kids, but meh. . . . I think that guys think girls are poo." In this respect, the in-game assistance was framed as "leveling the playing field." Despite what seemed at first to be a sphere that benefited women, the latent consequences of ample assistance, hand-outs, and superficial camaraderie kept some female gamers from being fully embraced in this male-dominated realm. However, the online gaming landscape has changed over time. As MMORPGs have flirted with the mainstream (spearheaded most notably by the explosive popularity of *WoW*, which has changed the player base considerably), gaming has begun to assimilate (albeit slowly) those who stretch beyond the aging stereotype of the male geek. Ivy noted her own experience changing as she continued to play *WoW*:

> I think it has gotten better over time, I remember it being a lot worse even two years ago, as the game gets more mainstream more girls play and I think that has changed it somewhat. I don't feel that I've ever been treated differently with regards to playing the game (i.e., no one has looked down on my playing skill or ability because of my gender), but in the socializing sense I think that males and females are treated differently, interacting with others in chat, making openly sexist remarks, some of that remains.

Some hyper-women stated that they felt less infantilized and sexualized by men indicating that, at least, some hyper-women had experienced a change in treatment over time (either their perception of the game worlds in general or their personal sense of acceptance). This is indicative of the slightly amorphous residue of online worlds' revolutionary potential, something that has escaped the full-crystallization of hyper-resonance and yet, may also indicate the closing door to a more post-gender reality.[34]

CONCLUSION

Hyper-women's experiences in game worlds are anchored by the hyper-resonance approach to play that requires women to prove they are women. Because men perceive women as possible romantic partners, women are required to prove their gender (which again, allows them to assume gender-sex congruence). Women may follow through to feel they are being properly recognized by gender, or to reap the benefits afforded them in the virtual world. In either case, men may treat women with suspicion only to then infantilize them (through virtual chivalry) once they prove to be solid women. By doing so, men seek to ensure that their affections embody proper (read: heterosexual) desire. "The desire for gender is not just the desire to conform and fit in, though that has a powerful effect, but an excitement felt as sexuality in a male supremacist culture which eroticises [*sic*] male dominance and female submission."[35] This expression is a historical project that has constructed dominance and submission as idealized forms for "heterosexual desire." Women are thus pushed into forms of femininity that adhere to the existing power structure within the games, which in turn informs us about the solid social world and what is framed as desirable. Femininity that is desirable is submissive, other-object, and female-bodied. This has implications for relationships outside the game, as well as future virtual relationships, because it may contribute to the reproduction and persistence of gender standards.

Women's reliance and men's enforcement of a "sexuality of male supremacy" reproduces normative standards of femininity and masculinity. This dysfunctional masculinity requires the fusion of a woman/object with a need to maintain a separate identity and space in which men have undisputed reign.[36] Women are thus infantilized and hypersexualized while men maintain a space of dominance, a common practice of a culture of patriarchy. In its most base form, the persistence of these standards has turned "playing games with women" into product for men's consumption.[37] Implicit in this vetting process, however, is the reality that online relationships (and ideally, any relationship) *can* circumvent gender and perhaps more importantly gendered power. However, desire is largely a gamble predicated on the assumption of congruence between gender and sex and this becomes apparent in MMORPG romances. This buttresses the importance of the inclusion of sexuality in furthering an equality imperative. Although the limitations of a sexuality-centered emancipation have been highlighted, at least some of the problem lay in desire that pushes men to create distance and objectify women when there could be equality and "power-with" sexuality, as well as desire in which women internalize their value in relation to men and as their capacity to be objects of desire and not partners.[38]

The preferential treatment of women gamers by men demonstrates that men and women are indeed perceived differently in MMORPGs. Although these "benefits" may be interpreted negatively, in-game success is often predicated on accomplishing tasks that require help from others, in-game currency, "epic" equipment, and powerful allies. Most players want nothing more than to be best. They want to have the best equipment, the best guilds, the best battles, and they can accomplish this more easily by proving to be solid women, which in turn casts them as "inferior" players and secures their access to this "preferential" treatment. These benefits change the experiences, roles, and difficulty of gaming online. Women are helped more, "handed gold like trash," and invited to battles and events more often. This contributes to women accruing status and success within the virtual world and arguably entices women to explore more gaming experiences, platforms, and spaces. Conversely, being awarded these benefits marginalizes women. Women are stereotyped as "gold-diggers," and are thus treated more as potential partners for men to woo with gifts and assistance than as *actual players*. This consequence is not founded on individual gaming ability, but on gender attributes alone. This has been illustrated in other spheres of everyday life where men misperceive their privilege as competence (or are misperceived by others) and thus continue to subordinate women.[39] The stereotype that gaming is a male-domain reinforces the belief that women cannot play, or cannot play well. This, in turn, contributes to men assisting women ingame. The assumed difference is then recreated and marshaled as evidence to deny women their identity as real gamers. However, the continued success and enjoyment women experience in virtual worlds is important for changing the practices that reproduce technology and gaming as a male-arena. If women want to play and enjoy playing, more games and spaces will be made for women. This is not to argue that women and men desire significantly different gaming and virtual world experiences, but to highlight that these spaces have embodied male-fantasies and male-standards. The call for the evolution of game worlds beckons both interactional relationships between players as well as institutional (game creators and companies) actors reopening the door to a virtual revolution before the crystallization of gendered stereotypes and behaviors is so deeply woven into the ethereal realm that those who inhabit its virtual walls will begin to believe the social arrangements within are simply the way things are and the way they should remain, dismissing the way they could be.

NOTES

1. "Author's field notes" are extracted from one of the three field note files (Empirical, Analytical, Epistemological) based on the tenants of John Lofland, David Snow, Leon Anderson, and Lyn H. Lofland, *Analyzing Social Settings: A Guide to Qualitative Observation and Analysis* (Belmont, CA: Wadsworth, 2006). Dialog and chat in-game (in situ interviewing, events) are chronicled in the empirical file. My critical musings and the cultivation of my theoretical nexus are within the analytical file. Finally, my difficulties, conflicts, and the shortcomings of my research approach are within the epistemological file.

2. *Lunar: Silver Star Story* released stateside in May 1999 for the Sony Playstation. This classic RPG has undergone several updates and re-releases and is generally touted as a timeless classic within the genre. The story elements involving Luna provided the foundation for the goddess paradox concept.

3. MMORPG is short for Massive Multiplayer Online Role Playing Game. Most use MMORPG and MMO interchangeably. However, I feel the RPG aspect of these games is important to stress as it relates to a specific genre that changes the nature of these games. I draw this distinction because a game like *Madden 2010* (a football game) or *Halo 3* (a first-person shooter, FPS) can be considered a "MMO" but both lack the role-playing elements that make MMORPGs a considerably different playing experience.

4. Some research on masculinities has highlighted how men may feel powerless as individuals, but are powerful as a social category. See Rocco L. Capraro, "Why College Men Drink: Alcohol, Adventure, and the Paradox of Masculinity." In *Men's Lives*, ed. by Michael S. Kimmel and Michael A. Messner (Upper Saddle River, NJ: Allyn & Bacon, 2000), 182–195; Michael S. Kimmel, "Masculinity as Homophobia: Fear, Shame, and Silence in the Construction of Gender Identity." In *Privilege: A Reader,* ed. by Michael Kimmel and Abby Ferber (Cambridge, MA: Westview, 2003), 51–74. I use this example because the reverse seems to speak to women's experiences in games: individual feelings of empowerment, but infantilized and sexualized by men.

5. "Solid world" refers to the world outside the game. This term eliminates the deployment of the word "real," since the virtual world is just as "real" as the solid one. "Solid" was the term used by a guild I belonged to for six months.

6. Avatars or characters in MMORPGs are virtual graphical manifestations that players control to navigate the virtual world. Avatars in MMORPGs are programmed in ways that mimic gendered behaviors, bodies, voices, dress, and so on. Avatars in MUDs are text-based, meaning they are "descriptions" of a virtual persona (similar to "character sheets" in tabletopping).

7. T.L. Taylor, *Play Between Worlds: Exploring Online Game Culture* (Cambridge, MA: MIT Press, 2006), 93.

8. Tanner Higgin, "Blackless Fantasy: The Disappearance of Race in Massively Multiplayer Online Role-Playing Games," *Games and Culture 4*, 1(2009): 3–26.

9. L. Konzack, "Rhetorics of Computer and Video Game Researcher." In *The Player's Realm: Studies on the Culture of Video Games and Gaming,* ed. by J. P. Williams and J. H. Smith, (Jefferson, NC: MacFarland and Company Inc. Publishers, 2007), 110–130.

10. Lynn Cherny and Elizabeth Reba Weise, *Wired Women: Gender and New Realities in Cyberspace* (Seattle, WA: Seal Press, 1995), 207.

11. The "gender/sex option" for each game is executed similarly, and is not surprisingly limited to the female/male binary. However, in some cases, gender may be exclusive to race; an example being the female only "Mithra" from *FFXI*.

12. Joland Dovey and Helen W. Kennedy, "From Margin to Center: Biographies of Technicity and the Construction of Hegemonic Games Culture." In *The Players' Realm: Studies on the Culture of Video Games and Gaming,* ed. by J. P. Williams and J. H. Smith (Jefferson, NC: McFarland and Company Inc., Publishers, 2007), 131–153.

13. Judith Lorber, "Believing Is Seeing: Biology as Ideology," *Gender and Society 7*, 4(1993): 568–581.

14. Lori Kendall, "'Oh No! I'm a Nerd!': Hegemonic Masculinity on an Online Forum," *Gender and Society 14*, 2(2003): 256–274.

15. Candance West and Don H. Zimmerman, "Doing Gender," *Gender and Society 1*, 2(1987): 125–151.

16. James W. Messerschmidt, "Goodbye to the Sex-Gender Distinction, Hello to Embodied Gender: On Masculinities, Bodies, and Violence." In *Sex, Gender, and Sexuality: The New Basics*, ed. by Abby Ferber, Kimberly Holcomb, and Tre Wentling (New York: Oxford University Press, 2007), 2–35.

17. Put simply the goddess paradox is a conceptual tool to highlight the dis/empowerment juxtaposition experienced by hyper-women once that had proved their status as solid women. I found it convenient that the Lunar narrative provided the imagery for this concept in that Luna was a being given seemingly infinite options and power, but who was also constructed as nothing more than a helpless girl.

18. Sheila Jeffreys, "Heterosexuality and the Desire for Gender." In *Theorizing Heterosexuality*, ed. by Diane Richardson (London: Open University Press, 1996), 75–90.

19. Days of play literally means days. MMORPGs calculate total play time in days, hours, minutes, and seconds. When I state that I played 500 days (across all the MMORPGs I experienced), I literally mean I was logged into the games for that amount of time. Interviewees reported anywhere from 20 to 650 days of logged play time (although for most their game time was ongoing).

20. Robert M. Emerson, *Contemporary Field Research: Perspectives and Formulations* (Long Grove, IL: Waveland Press, 2001).

21. Lofland et al., *Analyzing Social Settings*, 87.

22. Annette N. Markham, *Life Online, Researching Real Experiences in Virtual Space* (Walnut Creek: AltaMira Press, 1998), 71.

23. S. Kvale, *InterViews: An Introduction to Qualitative Research Interviewing* (Thousand Oaks, CA: Sage, 1996), 132.

24. Kathy Charmaz, "Shifting the Grounds: Constructivist Grounded Theory Methods for the 21st Century." In *Developing Grounded Theory: The Second Generation*, ed. by Janice M. Morse, Phyllis Noerager Stern, Juliet M. Corbin, Barbara Bowers, Adele E. Clarke, and Kathy Charmaz (Walnut Creek, CA: Left Coast Press, 2008), 127–154.

25. A. Coffey and P. Atkinson, *Making Sense of Qualitative Data* (Thousand Oaks, CA: Sage, 1996), 30.

26. I use the term "world resonance" to discuss the overlap between the solid and the virtual. The higher the resonance, the greater the overlap (or the less disturbance). Approaches to play that attempt to compartmentalize worlds have less world resonance and would accept more experimentation and flexibility.

27. Hyper-resonance is the encouraged approach to playing online that has manifested in the wake of a more mainstream player base. Hypers tout authenticity, honesty, and disclosure as the merits of this style of play.

28. All names I use are pseudonyms. These pseudonyms neither reflect the players' names, their avatars' names, nor their screen names. This ensures their confidentiality in the wake of systems like *WoW*'s "Armory" or *FFXIV*'s "Lode Stone" which allow anyone to search for any avatar by name alone. Pseudonyms chosen are generic fantasy genre names and are my personal preference.

29. "SM" (solid man) and "SW" (solid woman) indicate the players' solid world gender while "VW" (virtual woman) and "VM" (virtual man) indicate the gender of the avatar controlled by that player.

30. Some servers or guilds adhere to "role-playing" while playing the game. This is similar to "being in character" as with tabletopping and Live Action Role Playing (LARPing). Authenticity of solid-to-virtual identity is not stressed in these situations because all players understand that RP identities are theatrical. Here, there is less world resonance.

31. Allan G. Johnson, *The Forest and the Trees: Sociology as Life, Practice and Promise* (Philadelphia: Temple University Press, 1997).

32. Zek Cypress Valkyrie, "Cybersexuality in MMORPGs: Virtual Sexual Revolution Untapped," *Men and Masculinities*, 2010 (In press, currently available online at: http://jmm.sagepub.com/content/early/2010/04/23/1097184X10363256.full.pdf+html).

33. This was different from "mono-class" games (for example, *WoW*) where an avatar had only one class and, in order to play multiple classes, a player would have had to create multiple avatars. Players in multi-class games, by design, spend more time with less avatars (and have less "alts") than those playing mono-class games. It could be argued that because of this multi-class games encourage more investment (and therefore more reputation maintenance) than mono-class games where "alt" avatars are more common.

34. Donna Haraway, "A Cyborg Manifesto: Science, Technology, and Socialist-Feminism in the Late Twentieth Century." In *Simians, Cyborgs and Women: The Reinvention of Nature*, ed. by Donna Haraway (New York: Routledge, 1991), 149–181.

35. Sheila Jeffreys, "Heterosexuality and the Desire for Gender." In *Theorizing Heterosexuality*, ed. by Diane Richardson (London: Open University Press, 1996), 75–90.

36. Wendy Hollway, "Women's Power in Heterosexual Sex," *Women's Studies International Forum 7*, 1(1984): 63–68.

37. PlayWithMe is one example of a growing market that attempts to sell video game "play time" with women, accessed at http://playwithme.com/go/p119.submad_11844_1_3231.

38. Joseph Weinberg and Michael Bierbaum, "Conversations of Consent: Sexual Intimacy without Sexual Assault." In *Transforming a Rape Culture*, edited by Emilie Buchwald, Pamela Fletcher, and Martha Roth (Minneapolis, MN: Milkweed, 1993), 87–100.

39. Joan Acker, "Hierarchies, Jobs, Bodies: A Theory of Gendered Organizations," *Gender & Society 4*, (1990): 139–158; Deborah L. Rhode, *Speaking of Sex: The Denial of Gender Inequality* (Cambridge, MA: Harvard University Press, 1997).

Part III

Game Fans Speak Out

Chapter Ten

World of Warcraft and "the World of Science"

Ludic Play in an Online Affinity Space

Sean C. Duncan

Research into the dynamics of "play" in online games has understandably focused on the in-game collaborative and competitive activities that occur within the spaces of gaming activity. In Massively Multiplayer Online Games (MMOGs), this has included performing a role within a large, multi-player "raid" or shaping and crafting one's "avatar," while this has also included other game genres such as first person shooters (FPSs), forms of conversation, and interaction that takes place between players co-located within these spaces.[1] As research into the increased activity within these games has shown, there is much to be learned through a focus on how players represent themselves in such environments, how they interact with others, and how the constraints of a virtual domain can shape and influence their activities within them. Gaming in virtual worlds involves many forms of "play," yes, but also concomitant modes of interaction, the uses of virtual tools, the transfer of virtual goods, and other socio-cultural phenomena of interest.

However, as games studies—and the study of virtual worlds—has matured, the lines between what has traditionally been considered the domain of play within the game and the social/cultural practices outside of the game have blurred. Huizinga conceived of games as demarcating a "magic circle" into which external social and cultural phenomena could not enter;[2] this has been recently challenged,[3] raising the question of how best to study the means by which play in a virtual space extends and bridges into other forms of interaction. The impact of the "outside" world upon gaming has both called into question what "games" *are* beyond formal rule systems, as well as how other forms of "play"—including those found in academic pursuits and creative endeavors—might be relevant to understanding practices "around" virtual worlds.

Gee's notion of the "affinity space"—a fan "space," often presented online, in which players interact and develop an affinity for the common interest as well as one another[4]—has been used within learning and literacy literatures to help describe some of these extra-game play activities as well as other fan communities around media, such as *anime* fan fiction websites.[5] The notion of the affinity space is one in which fans and everyday players of games exhibit a number of common attributes—shared, common endeavors are worked toward, common virtual spaces are shared, new content is generated through activity within the space, the development of new knowledge (individual and distributed) is encouraged, and there are multiple routes to participation (among other features).[6] Gee painted a picture of the affinity spaces around several commercial videogames, highlighting how the online communities that form within these spaces can help to drive sophisticated learning and literacy practices within them. As these *ad hoc* "communities of practice"[7] have become increasingly prominent around games and virtual worlds, learning and literacy research with games has shifted toward understanding communities of gamers and the activities they partake of in addition to overt gameplay (e.g., Squire and Giovanetto's analysis of fan-made instructional sites,[8] or Gee and Hayes's look at "soft" or socio-technical modding activity).[9]

And yet, while Gee's notion of the affinity space holds much allure for those interested in further understanding how play moves both inside and outside of the confined "magic circle" of the virtual world, it is still a concept in need of further analysis. In this chapter, I aim to better analyze the means by which participants within MMOG affinity spaces make meaning of their own activities, and the discursive practices they employ in engaging with others "around" a particular game. I explore the notion that understanding "play" within virtual worlds *necessitates* a consideration of practices in the online communities and spaces around a game, and how players conceive of their activity within these spaces may be important—especially in how the communal activities in such spaces may complement or clash with the designers' intentions for providing these spaces to the gaming populace. That is, the analysis of the affinity spaces around virtual worlds can help us to both understand how play activities extend beyond the virtual space itself, as well as further examining the tensions that might arise between player activities and the constraints of such spaces.

Here, I focus on one such productive affinity space—the official online forums for the MMOG *World of Warcraft*, first published in 2004. In order to investigate the ways that play activities extend beyond the confines of the game itself, I focus on the *design-like* activities of participants within the online forums, looking at how players of *World of Warcraft* move from conceiving of their own play as disconnected from "everyday life" toward conceptions in which the "play" of the online discussion forum takes on the character of other, privileged activities, such as science and software design. How do players argue for redesigns of the game's rules and mechanics, and what ways of characterizing these design practices can we use to better understand their arguments? Additionally, how does participants' "talk" about these activities within the affinity spaces around *World of Warcraft* illustrate gaps in the conceptions of the spaces' purpose between the game's players and game designers (or their representatives)?

I use a discourse analysis approach[10] in order to highlight both the design-like activities within the online affinity space around *World of Warcraft*, as well as using it as an effective tool for characterizing the meaning-making activities of the affinity space's participants. Through the principled "un-packing" of discursive moves within a short exchange between one player and the game's lead designer (found within text in the online discussion forum for *World of Warcraft*), I aim to tease apart the means by which players and designers conceive of their activities within a selection of the discussions in this community.

CRAFTING TALENTS

World of Warcraft was, as of late 2009, still the most popular MMOG in history by far, with over 11 million active subscribers worldwide.[11] Featuring a "high fantasy" setting (with players adopting the roles of dwarves, elves, orcs, and other fantasy races), *World of Warcraft* has, in many regards, remained a very similar game experience since the game's launch—players connect to the game via personal computers, engage in play with one another and designed elements within the virtual world to progress through "levels" of personal game character development, complete joint collaborative and competitive "quests," and engage within complex financial activities. In 2010, as in 2004, players inhabit the virtual world of Azeroth with a virtual character of a specific "class" (a character type with unique abilities, such as Warrior, Priest, Druid, or Rogue), battling one's way through quests and monsters as a means of advancing one's character. With the achievement of higher and higher "levels" (a measure of character advancement, a staple of single-player and massively multiplayer role playing games), players activity becomes less about encountering new virtual locales, facing new monsters and quests, but instead in engaging within complex group play with other participants in the virtual world (raiding, player-vs.-player combat, participating in arena teams, and the like). The game's systems are numerous, and nearly all are tied to multiple other designed elements of the game—character level, quests, personal character attributes, roles within group play, and character class.

One of the most compelling features to high-end players of the game is the complex interconnectedness of the game's systems. *World of Warcraft* features hundreds of items of "gear," or virtual armor and weaponry that can change a character's abilities when used. Additionally, with every stage between level 10 and 80 (the highest player level currently in the game), players accrue a "talent point" for each, which can be spent by the player to customize his or her gameplay. For example, if the player of a Priest wishes to cause more damage against enemies rather than heal friendly characters, he or she can devote talent points to the abilities in his or her talent tree that support that desired mode of play ("speccing" as a Shadow Priest to cause damage rather than as a Holy Priest which heals). Steinkuehler and Duncan [12] noted that, as players gain a significant number of talent points and advance their character into higher and more complex forms of play, decisions regarding talent point allocation (creating viable "talent specs" or "talent builds") become increasingly important, especially with regard to managing these choices with respect to the kinds (and quality) of the gear available to a player. This has led, at times, to fans creating sophisticated computational and mathematical models of the game's systems to discern useful courses of action for their characters. And, as more players have reached the "level cap" (the highest level attainable within the game, currently level 80), they have turned to one another to evaluate what works and what doesn't with the current design of the game's talents, gear, prescribed activities, and social communities.

Coupled with other systems in the game—the aforementioned in-game gear, the specific class of character a player plays, and even the requirements of social groups (e.g., a player may wish to play a "tanking" Druid character while other players need the player to be a "healing" Druid character), the allocation of talent points has become an extraordinarily important aspect of the game. Being that players of *World of Warcraft* often play through a number of experiences on the way to the level cap, and, I suggest, face these social play concerns along the way, determining not just the best talent spec to choose but arguing for *new* configurations (or, as I label them, *redesigns*) of the game's many systems dominates some players' activities "around" the game. The process of redesigning has become such an important aspect for some players that, in the social spaces around the game, many have argued ways to modify the game's mechanics toward desired ends. In other words, players have begun to enact a recombinatory play activity with the game's existing mechanics—I argue this is a decidedly *ludic* form of play, as opposed to one focused on, say, expanding the game's narrative elements, and indicates that many dedicated *World of Warcraft* players have taken up the task of attempting to determine the relationships between game systems, and the ways that they should change.

Though players have apparently little to no impact upon the game's over-arching narrative (planned and written by the game's designers on a roughly two year schedule with periodic updates in-between), they do have the facility to "build" their characters' abilities to fit their particular style of play, then test and evaluate these choices. Given the iterative nature of design, some players will use tools to collect statistics on their gameplay under certain gear and talent allocation conditions, then report their results back to the community for evaluation. Dubbed "theorycrafting," the collection of data on gameplay choices and the development of mathematical models of the game's systems has become a productive activity for a relatively small number of dedicated players, as well as a resource to consult for a great many other players. Therefore, it is unsurprising that discussions of these redesign proposals and their broader implications dominate sections of the online forums provided by Activision Blizzard, and are used by players in order to make clear their preferred redesigns to the game's designers.

Steinkuehler and Duncan illustrated that a variety of sophisticated discursive practices and informal scientific reasoning practices were present in the online discussions over talent "builds" and talent point allocation in the official *World of Warcraft* class forums (http://forums.worldofwarcraft.com). However, we did not evaluate the role(s) that game designer representatives—employees of Activision Blizzard itself, or player community managers, both indicated by "blue posts" (posts highlighted in blue text)—played in these discussions. I hypothesize that, as this game has quickly moved through two major expansions, many small rollouts of new content ("patches"), and the fixing of several smaller bugs ("hotfixes"), the nature of the player interaction with "blue posters" has become a significant factor in shaping ludic play within the online affinity spaces around the game. These representatives of the company have designed (and continue to alter) the game's underlying rules, occasionally entering into the affinity space to explain changes to a particular character class. As of 2009, "blue posters" have frequently entered into dialogue with players about changes they have made to the classes, evaluating player proposals for redesigning the game's mechanics, and serving to explicate aspects of the game's continual design process to the paying customers of *World of Warcraft*.

I suggest that many of the *World of Warcraft* forums feature player participation of this ludic variety—play that involves retooling the game itself, proposing changes to the game's mechanics, and testing their viability. These player proposals for changes to the game's underlying systems are done within the presence of professional game designers (or, at least, their representatives, such as community managers) and presents a unique case in which an 11 million player base has easy access to the game's designers for their proposals to (at least potentially) be heard. Player participants within these threads may thus be emboldened to provide more complex and sophisticated arguments toward the evaluation of *World of Warcraft*'s continual redesign process than is typically thought to be within the realm of online gamer discussions; this case may show that the hypothesized design activity can directly impact the *player's* "play" activities. For example, it is relatively common to witness a proposal to change a class which a player has invested a great deal of time and energy in, but is perceived as being "broken" by that player and its redesign is taken up as a project to work through within the affinity space, as a means to improve the class for an individual player's gameplay goals, but also to engage within a larger community.

Thus, my concern is with the activities and practices of players within the affinity space and I will stay relatively agnostic toward the content of the specific proposals suggested by players. Instead of evaluating the quality of their redesign proposals, I am instead looking past the specific redesign arguments toward characterizing the ways that these fans utilize *discursive* practices in service of their goals. Therefore, I will not seek to assess whether or not, say, a specific gear change is truly necessary to promote balance between game systems, but will instead analyze how players and designers have argued these redesign proposals through discursive means.

THE OFFICIAL DISCUSSION FORUMS

For Steinkuehler and Duncan, talent spec discussions were characterized by investigating a specific class forum (the Priest forum). However, a number of subtle changes to the game since that study's data was collected (in November 2006) have caused a reconsideration of the most productive locus for analysis. Most significantly, the game's mechanics have changed in ways too numerous to fully catalog in this chapter, including new character races (Blood Elves and Draenei) being brought into the game through an expansion, alterations to character classes associated with only one "faction" (large scale team of players within the game, Alliance and Horde) being now imported to the other faction (Draenei Shamans and Blood Elf Paladins), as well as new content areas (Outland, Northrend), numerous new quests, and new "instances" (areas for "raiding," or group play versus the game's environment or other players).

But, most importantly for understanding ludic play within *World of Warcraft*, the game's mechanics have shifted in a common manner for many different classes of characters. As the game has expanded, mathematical relationships between mechanics have changed, and the level cap has increased, more talents have emerged that provide players with new abilities, spells, and ways to configure one's character to facilitate a variety of play styles for each character class. This ongoing process of design has in some ways blurred the lines between different character classes, providing a number of character classes with multiple forms of play to enact in different areas of the game (at least more so than in the original game's release in 2004). For instance, characters within the Druid class have always been configurable as a "tanking" class (to take damage and hold the attention of the monster during group play), "DPS" class (damage per second, or damage-dealing at the monster from either close by or afar), or "healer" class (focus on keeping the tank alive and, to a lesser extent, other members of the party). But, by 2009, the viability of certain "specs" had changed with the content of the game—for instance, while a tanking-specced Paladin may have been some-what viable in the past, players with this class now often serve as healers in high-level group play.

Activision Blizzard created a new set of "Class Role" forums in late 2008 in order to address the common activities that cut across a number of player classes. The three in existence at the time of this chapter's writing are "Tank-ing," "Healing," and "Damage Dealing," allowing for players of several classes to interact with one another and the Activision Blizzard employees (or community managers) populating the forum's threads. Given the wide variety of player classes that can potentially perform damage-dealing roles in *World of Warcraft*—Rogues, Warlocks, Druids, Mages, Priests, Hunters, Shamans, Paladins, and others—this forum has become quite popular in a very short period of time. Undoubtedly adding to the forum's popularity has been (through September 2009) the very active involvement of *World of Warcraft*'s lead game designer, Dr. Greg Street (known online as "Ghost-crawler").

Until quite recently, Ghostcrawler had been the driving force of many of the discussions within the Damage Dealing forum. With a Ph.D. in marine science from the University of Texas at Austin, former experience as a uni-versity assistant professor (at the University of South Carolina), and game designer for other gaming properties (including design experience on Ensem-ble Studios' *Age of Empires*), Ghostcrawler has brought a number of forms of expertise to his role as lead game designer for *World of Warcraft*. Until his recent "forum retirement" (in September 2009),[13] Ghostcrawler was respon-sible for both answering many questions about the game's current and future design as well as moderating forum behavior (through declarations of good forum behavior and the banning of many players). Having at least the illusion of direct access to the lead game designer for a shared virtual world of over 11 million players was unprecedented for contemporary gaming affinity spaces, and understanding the ways that players employ discursive and de-sign practices in their conversations with Ghostcrawler (and the other "blue posters") is, I argue, an important case for understanding these deeply *ludic* forms of play, especially if they may have consequences that could affect the gameplay of millions of players.

In the next section, I present a single exchange within one of these threads—a post by Ghostcrawler, reacted to by a single player. Through an investigation of the specific ways both participants presented their assumptions about the forms of "play" present within the affinity space, I have conducted a stanza-based Discourse analysis. I argue that assumptions about what the activity of the affinity space is *for* are uncovered through this analysis and, by extension, may characterize many other players as well as prevailing attitudes of the Activision Blizzard design team.

DISCOURSE ANALYSIS

The exchange analyzed in this section is from a series of posts begun in April 2009 on the Damage Dealing forums within a thread entitled "Conflag changes on top of immolate?" The context of the discussion was about "Conflagrate," a spell for Warlocks near the middle of their "Destruction" talent tree—an ability to cause a specific type of damage ("fire") to opponents. Viewed by some as a critical spell for Warlocks who have chosen to follow the Destruction tree (a high-damage dealing talent spec featuring direct damage rather than damage over time, as seen in the Warlock's "Affliction" tree spells), the issue in this exchange was over player reactions to a recent "hotfix" performed by Activision Blizzard. A very short time after a previous change to the game mechanics, the designers opted to change the behavior of this spell with the net result of lowering Warlocks' damage-producing viability in certain in-game combat situations. Interested players (many who had vested interests in changes to the Warlock class) took to the forums to try to understand and debate the significance of these changes.

As a consequence of this hotfix, many of the Warlock-playing participants in this thread were unhappy with the change, and began to openly question the reasons for it. This went in several directions, with participants both suggesting changes to the specific degrees of damage certain spells would create, while others discussed changes to a "glyph" (an ability affecting other spells) that would radically alter how Conflagrate would behave in conjunction with other spells (Immolate, in particular, referred to in the thread title). However, as the debate continued, a number of issues arose questioning the need for these changes, involving discussions of the process that Activision Blizzard undertook in determining that the changes were warranted, and, in general, the role that player efforts to gather *data* on their activities should have had in helping Activision Blizzard make these decisions (if at all).

In particular, the contentious issue of a player "theorycrafting" tool called "SimulationCraft" (also referred to as "SimCraft" by some players) became discussed by both players and Ghostcrawler in this thread. This thread quickly turned into a discussion of the practice of high-end players to create elaborate and sophisticated computational models of the game's systems, based on empirical evidence from numerous players as well as certain assumptions about the game's mechanics. SimCraft, "a tool to explore combat mechanics," is an open-source, command-line program for capturing damage-dealing data in particular, seeking to simulate damage within a variety of contexts (both several kinds of group play and individual).[14] Since the true relationships between these elements of the game are unknown to anyone other than Activision Blizzard employees, there is debate within the gaming community as to the accuracy of these models, as well as their utility.

As the discussion of Conflagrate progressed, SimCraft data was used by participants in order to justify their arguments, often in a manner that referred to the SimCraft data as "the state of affairs." This clearly was not the interpretation of these data that Ghostcrawler preferred, who expressed his opinions on the use of such tools and the ways in which they informed the design processes of Activision Blizzard designers under the company's employ. He stated:

I've commented on Simcraft (and any similar tool before) but I'll repeat myself.

1. It's awesome to see players dedicating that much effort to *WoW*. It really is. They show a passion for the game and dedication to improving both the player's effectiveness and the game in general. It is humbling in a way.

2. Those tools are very difficult to make. I'll give a shout out to Toskk's Feral spreadsheet, which represents an enormous effort and is still being refined constantly. Getting that kind of accuracy and precision for every spec in the game is going to be challenging.

3. As the community continues to offer feedback, refine and grow to accept Simcraft (or any tool), so will we. We aren't going to spend a great deal of our effort to troubleshoot or verify their assumptions. They are third party tools.

4. At the end of the day, the Blizzard designers are going to balance the game. Not the community. Not Simcraft. Not any external tool. If you want to use those numbers as part of your argument, that's awesome. But just posting those numbers and saying "Fix it," isn't going to work. I've said this a lot lately, but you should stop approaching every

potential change as "What do we have to do to get you to make this change?" The answer is there is nothing you can do. You can give us information and we will use that information to make informed decisions. But we, not the community and not external tools, are going to make those decisions.

5. Do remember that statements such as "warlocks do 6800 dps" are virtually meaningless. The actual damage you can do on a raid boss varies enormously depending on your skill, gear, lag, luck, the boss encounter, and the buffs you have. Estimating a maximum theoretical dps is a very useful piece of data and nothing to be scared of. But that does not mean you will ever see those numbers with any consistency. Do not take them as gospel. Ever.

In a subsequent post, a player that I will refer to within this chapter as Nawaf posted a lengthy response. Nawaf, a player with a Gnome Warlock at the current level cap, quoted the entirety of Ghostcrawler's post above, and responded in kind with an evaluation of SimCraft's utility in assessing the issues with Conflagrate, but also addressing larger issues of process, design, and the presumed intent of the designer toward fostering a stronger sense of "competitive balance" within the game (which, in this thread, a number of Warlock players felt was being violated through the weakening—or "nerfing"—of the Conflagrate spell). Nawaf's post follows:

> The problem is deeper or shallower than you seem to comprehend—depending on your perspective. Here's what I mean (numbers below do not correlate to your listed numbers).
>
> 1. People perceive simcraft data as validation, much the same as people in my field (Quantum Optics) feel molecular dynamics validate their data. I'm not sure what kind of exposure you had to modeling in your marine biology PhD, but simulation data is often used and accepted in the academic community as a tool for understanding the underlying effects of individual variables. Yes we're talking about toy models. But the same can be said for a great many "real" experiments in science. Model systems are also toys. The benefit of studying toy models through simulation, as opposed to studying them in experiment is that you have so much more control over every possible variable. There's so much more data output that can be generated from simulations. The World of Warcraft really isn't that much different from the "World of Science." WWS and raid parses are similar to experiments performed on model systems. Simcraft data is analogous to molecular dynamics.
> 2. I understand that you're loth to use a piece of third party software. Frankly, though, the Simcraft program isn't a massive undertaking. You could hire a graduate student, such as myself, to code the sucker up for you over the summer. Hopefully you guys are already a step ahead of me on that.

3. I have to wonder what the true goal of PvE balance even is. Functionally, as long as a single spec exists that's competitive, one could be satisfied. It would make for a very inflexible and disappointing game, however. If you wish to balance towards having multiple, competitive specs, what is the purpose in that? Is the purpose to provide options to the community? That is my understanding of it, but I could be wrong. If you do want to give people options, then I'm going to tell you that you've got a tough row to hoe there. As you've noted, many people will just spec whatever is perceived to be the best spec. Why? Because they simply don't have time or desire to figure it out, wasting time, gold, and possibly raid performance in the process of experimentation. Do we want to punish these people because they are unwilling to devote their entire time to finding that one perfect spec they both love to play and performs well? I don't think so. This brings me to the conclusion that handling public perception of certain specs is a non-trivial part of your agenda, which would explain your vehement downplaying of simulationcraft.

In the remainder of this section, I will unpack this extraordinarily complex exchange for what Ghostcrawler and Nawaf's disagreement can tell us about ludic play within the affinity space, the social construction of knowledge in these forums, and the clashing interpretations of what the design process *should* be. For the purpose of clarity, I will first isolate key phrases from each poster's numbered responses rather than parse the entirety of these two posts into stanza form. It should be noted that both Ghostcrawler and Nawaf presented their posts in such a fashion—rather than present undifferentiated arguments, the *form* of both responses follows numbered lists, presumably intended to represent a structured argument (in Ghostcrawler's case), and/or potentially to differentiate separate points of concern (in Nawaf's case). Below, I have identified select phrases for further analysis, with portions italicized to highlight key moments in the discussion.

Ghostcrawler

1. "It's *awesome* to see players dedicating that much effort to *WoW*. It really is. They show *a passion for the game and dedication to improving both the player's effectiveness and the game in general.*"
2. "Those *tools* are very difficult to make. I'll give a shout out to *Toskk's Feral spreadsheet*, which represents an enormous effort and is still being refined constantly. *Getting that kind of accuracy and precision for every spec in the game is going to be challenging.*"
3. "We aren't going to spend a great deal of our effort to troubleshoot or verify their assumptions. *They are third party tools.*"
4. "*Blizzard designers are going to balance the game. Not the community. Not Simcraft. Not any external tool.* If you want to use those numbers as *part of your argument*, that's awesome. But just posting

those numbers and saying "Fix it," isn't going to work. I've said this a lot lately, but *you should stop approaching every potential change as "What do we have to do to get you to make this change?" The answer is there is nothing you can do. You can give us information and we will use that information to make informed decisions. But we, not the community and not external tools, are going to make those decisions."*

5. *"Estimating a maximum theoretical dps is a very useful piece of data* and nothing to be scared of. But that does not mean you will ever see those numbers with any consistency. *Do not take them as gospel. Ever."*

Nawaf

1. *"People perceive simcraft data as validation, much the same as people in my field (Quantum Optics) feel molecular dynamics validate their data.* I'm not sure what kind of exposure you had to modeling in *your marine biology PhD,* but *simulation data is often used and accepted in the academic community as a tool* for understanding the underlying effects of individual variables. Yes we're talking about toy models. But the same can be said for a great many "real" experiments in science. Model systems are also toys. *The benefit of studying toy models through simulation, as opposed to studying them in experiment is that you have so much more control over every possible variable. There's so much more data output that can be generated from simulations. The World of Warcraft really isn't that much different from the "World of Science."* WWS and *raid parses* are similar to *experiments performed on model systems. Simcraft data is analogous to molecular dynamics."*

2. "I understand that *you're loth to use a piece of third party software.* Frankly, though, the Simcraft program isn't a massive undertaking. You could *hire a graduate student, such as myself,* to code the sucker up for you over the summer. Hopefully you guys are already a step ahead of me on that."

3. "If you wish to balance towards having multiple, competitive specs, what is the purpose in that? *Is the purpose to provide options to the community?* . . . I don't think so. This brings me to the conclusion *that handling public perception of certain specs is a non-trivial part of your agenda, which would explain your vehement downplaying of simulationcraft."*

In Ghostcrawler's first numbered point, we see a framing of the use of simulation tools by the player community as "awesome," and representative of players' best intentions to improve both "the player's effectiveness and the game in general." Right off the bat, Ghostcrawler conveyed that he understood that this was a predominant concern of high-level *World of Warcraft* players, and lauded the dual goals of using tools such as SimCraft—improving individual performance and engaging within the forum communities to advocate desired design changes for the game. However, Ghostcrawler quickly began to differentiate their *use* from the systems that they were modeling—in point 2, he continued by describing them as "tools" and giving credit to specific quantitative tools that he deemed of value ("Toskk's Feral spreadsheet," a detailed computational model for Druid damage dealing calculation),[15] then he began to question the general strategy of simulation for every set of systems in *World of Warcraft* ("getting that kind of accuracy and precision for every spec in the game is going to be challenging").

Shifting into explicitly stating his position on SimCraft, Ghostcrawler labeled them "third party tools" (point 3), terminology that seems chosen to qualify the previous use of "tools" as something that was "third party," evocative of other products created by parties that are neither the primary developer, nor the end user. It is this specific phrasing that reflects Ghostcrawler's stance on SimCraft and similar methods of simulation—instead of arguing that these are "player tools" or "gamer tools," Ghostcrawler pushed simulation efforts like SimCraft out of the mainstream of player activity. Additionally, the choice of the term "third party" highlighted a particularly *industrial* framing of the game design and development process; Ghostcrawler emphasized a common term that is applicable to many forms of software and consumer electronics in general.

Furthermore, in the subsequent paragraph (point 4), Ghostcrawler continued and differentiated his conception of the development and use of these "third party tools" as something other than what players normally do in *World of Warcraft*. He stated, "Blizzard designers are going to balance the game. Not the community. Not Simcraft. Not any external tool," making a sharp distinction between the developers (Activision Blizzard), the community, SimCraft, and any "potential tool." For Ghostcrawler, these appear to be separate categories—the "community" and SimCraft (a tool used by and, here, standing in for a dedicated subset of players) were not the same thing, with a different set of concerns than (Activision) Blizzard. Further detailing the differences between SimCraft, the community, and the developers, Ghostcrawler continued with, "[Y]ou should stop approaching every potential change as 'What do we have to do to get you to make this change?,'" making a telling shift to the second person, and a prescriptive tone ("you *should* stop," emphasis mine). At this point, Ghostcrawler made clear the value he saw in the use of SimCraft (and, presumably, similar tools) by the community: "You can give us information and we will use that information to make informed decisions. But we, not the community and not external tools, are going to make those decisions." That is, the power resides in Activision Blizzard to make whatever changes they deem necessary, using the "information" provided by the community to make "informed decisions."

Through the course of just this brief post, Ghostcrawler moved from lauding the "passion" of the community in investigating the systems of the game to more clearly defining a process model of what the activity in the forum *should* be conceived as. By emphasizing that the community (and its tools, such as SimCraft) are creators of "information" which are used by the design team to make "informed decisions," Ghostcrawler outlined a stance on the epistemological project of redesign proposals within the Damage Dealing forums. He continued, stating (in point 5) that "Estimating a maximum theoretical dps is a very useful piece of data and nothing to be scared of"—casting the results of SimCraft and similar tools as "estimates," "theoretical," and "data," but exhorting *World of Warcraft* players to "[N]ot take them as gospel. Ever." Here, Ghostcrawler ended with as clear of a statement as seems possible regarding his opinions on the validity of results provided by the community; these results are *information* to be used in the process of design for *World of Warcraft* (an activity which Activision Blizzard employees solely engage in), and this information, while useful, is community-generated and may or may not imply courses of action that can sync with the (private) models that Activision Blizzard has available to them.

Thus, we see that Ghostcrawler's post serves to provide means of differentiating terminology vis-à-vis "third party tools" and the role of players as, essentially, providers of feedback, while also indicating a preferred way for players to interact within the affinity space. Note that in point 4, Ghostcrawler stated "you should stop approaching every potential change as 'What do we have to do to get you to make this change?'" As lead game designer and also as an active participant within the forum, this gives us a glimpse into Ghostcrawler's process of shaping the community practices within the online discussion forums. A large proportion of threads on the *World of Warcraft* forums were closed by "blue posters" (many by Ghostcrawler himself) for inappropriate content and, from some players' perspectives, for the persistent asking of questions that Ghostcrawler did not want asked. In point 4 from the post above, Ghostcrawler attempted to shape the community dialogue by prescribing specific forms of questioning: rather than asking Ghostcrawler what the community needed to do to effect a particular desired change, he presented a fatalistic (at least from the position of the player) perspective on the usefulness of these questions for effecting change: "[T]here is nothing you can do." The power to make changes to the game's systems resides solely in Ghostcrawler's and Activision Blizzard's hands.

For the purposes of this analysis, I have excluded other responses to Ghostcrawler's post (of which there were many), in order to focus on the post by Nawaf that appeared seventeen posts and 41 minutes later in the thread.

As with Ghostcrawler's post, the tenor of Nawaf's indicated that he or she too was dealing with the ways that the fan community and designers should shape the game's underlying mechanics. However, as with Ghostcrawler, Nawaf moved beyond these issues into discussing the process of design, the connections between design and scientific practice, and the underlying motivations of Activision Blizzard designers in participating within the online discussion forums. Perhaps taking up the change in the topic of the thread away from discussions of Conflagrate to the impact of Ghostcrawler's pronouncements, Nawaf provided a counterpoint to the game designer.

In his or her first point, Nawaf began with an explanation at the level of the community, not of an individual, stating that "People perceive simcraft data as validation, much the same as people in my field (Quantum Optics) feel molecular dynamics validate their data." It is notable that Nawaf explicitly spoke of "people," presumably the users of SimCraft data in the *World of Warcraft* community, but then very quickly flagged him or herself as someone with science expertise, using an analogy from quantum optics on the ways that physicists use "molecular dynamics [to] validate their data." With an appeal to Ghostcrawler's academic past ("your marine biology PhD"), Nawaf appears to have attempted to frame the discussion not as one about game *design* but about the similarity of gamers' practices and those in the academic communities that he or she presumably believed to be of value (and, which Ghostcrawler was once trained and employed within). Ghostcrawler's use of "tool" was picked up by Nawaf here, and subtly challenged—Nawaf stated "simulation data is often used and accepted in the academic community as tool," indicating, perhaps, that Ghostcrawler's conception of a "tool" in an industrial manner was not the only way to conceive of SimCraft. In other words, the meaning of the term "tool" became contested—instead of a tool serving as a vehicle to *information* for game designers, the tool was conceived as a device to build *simulations* useful to both designers and the community. A subtle shift, but one with profound implications.

In point 1, Nawaf continued by advocating the benefits of using simulations (or "toy models") in science. He or she argued that "The benefit of studying toy models through simulation, as opposed to studying them in experiment is that you have so much more control over every possible variable. There's so much more data output that can be generated from simulations." Evocatively, Nawaf further claimed that "The World of Warcraft really isn't that much different from the 'World of Science,'" continuing by linking "raid parses" to experimental uses of computational simulations in the sciences. "Simcraft data is analogous to molecular dynamics," for Nawaf and, clearly, the activities of dedicated users of simulation tools within the *World of Warcraft* player communities were intended by Nawaf to harken back to methodologies and community practices found in scientific settings.

Yet, Nawaf also appeared to be sensitive to Ghostcrawler's framing of these tools as "third party," picking up the terminology within point 2 by stating that Ghostcrawler was "loth [*sic*; loath] to use a piece of third party software," further arguing—rather ostentatiously—that a "graduate student, such as myself" could bring such endeavors in-house. What's most interesting here is that Nawaf seemed to recognize Ghostcrawler's reticence to cast the community design of "third party tools" as valid in his first point and made a rather lengthy defense of their use in the sciences, while in the second point brought it back to Ghostcrawler as (an admittedly brazen) proposal to incorporate these methods into Activision Blizzard's formal design process. This opens the question of what exactly about the appeals to science Nawaf found most important: Was it the use of the methodology of the "toy model" of the simulation, or was it his or her emphasis that an epistemological stance ("the *World of Warcraft* really isn't that much different from the 'World of Science'") undergirds the use of these simulations within scientific *communities*? This remains unresolved, but Nawaf's third point does call into question his or her interpretation of Ghostcrawler's intentions by engaging with players within the affinity space at all.

In his or her last point, Nawaf highlighted that the issue of "competitive balance"—the issue that brought many players of Warlocks to the thread in the first place—is a contentious one. Asking "Is the purpose to provide options to the community?" Nawaf questioned the *goals* of the redesign that had been the underlying issue discussed throughout the thread, and continued by asserting "that handling public perception of certain specs is a non-trivial part of your agenda, which would explain your vehement downplaying of simulationcraft." For Nawaf, then, Ghostcrawler's "downplaying" of tools such as SimCraft had a much larger purpose than just communicating his model of the design process to the community, and also reflected his intent in participating within these threads. Taking Nawaf's comment at face value, it would seem that at least some of the participants in the forum did not view Ghostcrawler as operating without a public relations agenda through his involvement in this particular affinity space. At the very least, the devaluing of community-designed and community-interpreted tools such as SimCraft reflects, perhaps, the realities Ghostcrawler faced in determining how to manage redesigns of the game *and* the shaping of the public perception of such changes, which is certainly an important concern for a company attempting to keep millions of players subscribed to their service.

Therefore, this Discourse analysis reveals that there were two conflicting framings of what the discussions in the affinity space were actually useful *for*, as well as two different implicit arguments on the role that community tools (and, by extension, community redesign proposals) should take within these forums. In the next section, I step back from this exchange and attempt to evaluate the conflict between interpretations of this affinity space, as well as explore what both participants' framing narratives might mean for the understanding of play activities within affinity spaces around games such as *World of Warcraft*.

PLAY BETWEEN SCIENCE AND GAME DESIGN

First, it is clear through this exchange and analysis that there were various competing conceptions of "design" within this space, with Ghostcrawler's understandable focus on the professional, commercial practice of game design being at odds with the community-based, fan-oriented, science-like conception forwarded by Nawaf. Additionally, it is telling how the influence of "blue posters" (or *the* most important blue poster, Ghostcrawler) shaped the discourse within the thread. By stepping in and raising concerns as to how and why participants in the affinity space should be using specific tools available to them, Ghostcrawler at once drew the conversation away from the specific redesign proposals being discussed as well as created a focal point for the community within the space to rally for or against. The presence of the game's lead designer was thus significant for both the "official" declaration of what the community practices advocated by posters such as Nawaf were for, as well as for an exhibition of the kinds of power that the game designer can have in directing the flow of the discourse within these kinds of affinity spaces.

Two primary framings were at play within the space—the *World of Warcraft* forum as a "feedback space" to inform the Ghostcrawler's goals of assessing information from the community, and the productive "science-like" space advocated by Nawaf, in which players have a hand in making meaning which, they hope, will translate into design changes for the game. These are not necessarily incommensurate framings of the collective activity of players within the affinity space; clearly, Ghostcrawler intended for there to be active discussion within the forums, and promoted the judicious use of player-created "third party tools" such as SimCraft. Similarly, Nawaf suggested that these activities should be brought "in house" to better support the designers' goals of basing their design decisions upon actual player data. What was under debate, however, was reflected within Nawaf's emphasis on the *academic community*—by making an analogy to forms of intellectual discourse, community standards of evaluation, reasoning with data, and the sophisticated use of computer simulations, Nawaf was arguing for players to be seen as more than just feedback providers, and was, instead, arguing to bring the community into the design process in some acknowledged fashion. In other words, Nawaf was fighting for legitimacy for the ludic play activities in the gaming affinity space, and sought for the game's lead designer to acknowledge the intentions of him- or herself, as well as SimCraft users in general.

This framing of the community as an academic one implies a conception of a form of ludic play that is necessarily social, constructive, and evaluated at a community level—not simply by a single designer (or set of designers) holding all the power. While it seems unlikely that Nawaf did not understand that this was in fact Ghostcrawler's role, it is still telling that Nawaf appealed to Ghostcrawler's previous experiences as an academic scientist. Nawaf's appeals to analogies from physics and, in general, his or her attempts to engage with Ghostcrawler by referring to his scientific past were an interesting and not uncommon move within these forum debates. In a different exchange within the Damage Dealing forums, a similar (and arguably more congenial) attempt drew out a response from Ghostcrawler on how he saw his scientific expertise impacting the process of game design. All that remains of that thread is the first post and Ghostcrawler's later response, as archived by the MMO-Champion "blue tracker."[16] The thread was originally posted in the Damage Dealing forums in November 2008 by a poster referred to here as Namorada, and has been edited down for space concerns:

> I'm currently working on my PhD in physics. Through the magic of the internet we all know you have a PhD in marine science. I was just wondering if you felt like being a developer for *WoW* was similar to your experience doing research in graduate school? I've always felt that, being a research oriented person, I would make a good developer (no, I'm not looking for a job!). Here are some parallels I see:
>
> You guys look at the current evidence you have and develop a theory. Then you carry out controlled experiments to validate your theory. Considering you can never 100% simulate a real world *WoW* environment your experiments correctly validate or disprove your theory most of the time, but not all the time. I guess that's more akin to engineering type research.

Several hours later, Ghostcrawler responded (also trimmed for space, with sections italicized for emphasis):

> A lot of what we do is try to change relationships in very specific ways without causing unwanted effects. Ultimately that means *understanding how different parts of a complex system work together.* In that way it is very much like doing a controlled experiment (in the hypothesis-testing sense).
>
> A lot of what we deal with is probability and statistics. *You have to isolate random effects from actual statistical relationships. You have to notice trends within noisy data.* You have to understand that when you see X and Y together that you cannot posit that one causes the other. You have to understand the value of large sample sizes and be hesitant to draw conclusions from small

sample sizes. There is a lot of math involved, and the math can be staggeringly complex. In my experience, statistics is woefully misunderstood by a lot of people.

Because I did a lot of field work, I feel like *I have reasonable appreciation for the difference between theoretical predictions and real world measurements.* Many players (and quite a few game developers) accept at face value predictions that may not be reproducible in the actual game. This is why I always warn everyone not to read too much into theorycrafting—you don't always know the assumptions and errors that exist in those estimations, and often trying to replicate those results in the chaotic, ever-changing, random and potentially laggy game can be challenging to say the least.

Again, *one of the best uses of theorycrafting is to learn how the relationships work.* "Wow, look how my hit change affects dps more than crit, AP or almost any other variable." That's a good result. You learned something. It's a big leap from there to "Our simulation shows that spec X will dominate and spec Y will always fail."

In this evocative passage, we see Ghostcrawler's intent, as a designer, to carry through with scientific approaches to data interpretation, understanding complex systems, separating causation from correlation, and suggesting the limits of mathematical modeling for understanding the systems of the game. Here, it is clear that Ghostcrawler "gets" the desire to understand the systems of the game, and even the motivating power of learning within these spaces to drive continued engagement with the game ("That's a good result. You learned something"). Much like the work that was found within communities of passionate *World of Warcraft* players in Steinkuehler and Duncan's work, Ghostcrawler's post indicates that the designers of the game have openly advocated for the employment of scientific skills and dispositions as a key part of discerning how the game "works," and that fostering a community-level sense of "play" centered on learning and understanding the game's systems was useful.

Yet, like the exchange with Nawaf, note that Ghostcrawler stopped short of advocating that these fan community forms of ludic play are a part of the *design process*, though Namorada's original post clearly implied some form of transfer between the two sets of skills ("I've always felt that, being a research oriented person, I would make a good developer"). Ghostcrawler's continued framing of the role of player investigations into the game was at the level of providing feedback for developers, but not straying too far into the realm of design proposals. From Ghostcrawler's perspective, it appears that the player community's activities in this regard are ultimately always in service of where power is ultimately consolidated: the Activision Blizzard employees who, with access to different tools and the game's underlying code, are best able to evaluate the empirical results produced by the community. These forms of play are useful for keeping players interested in the game, but participants should not believe that this brings them into being a part of the *actual* design process (other than as providers of "feedback").

At the same time, posters such as Nawaf and Namorada expressed seeing a similitude between the practices of science and game design—the play activities present within the affinity space were not significantly different from those that practicing game designers performed in the act of designing (and redesigning) a game. While Nawaf advocated an alternate framing of the activity within the affinity space, both the players and the designer present in these data indicated a commonality between practices, just a very different emphasis in terms of the *value* of these practices. By highlighting the player-created tools and player-created knowledge as "third party," Ghostcrawler was at once both qualifying the results as potentially containing "assumptions and errors," but also seeking to limit the players' inferences made using such tools. That is, while the academic community model favored by Nawaf may contain evaluation and assessment of the game's systems at the community level, for Ghostcrawler, the very real factors of being employed to maintain a multi-billion dollar commercial enterprise suggest that this will never take place. To be blunt, it is Ghostcrawler's *job* to be the game's lead designer, and thus he is forced to walk a fine line between encouraging productive game design-like and science-like practices within the fan community—not to mention curtailing such activities when they become counter-productive (either for his time or in propagating misconceptions about the systems of the game within the community). The role of the "play" within this gaming affinity space is necessarily curtailed by the realities of intellectual property, the management of a multi-billion dollar gaming franchise, and the individual desires of those under Activision Blizzard's employ.

In this way, we can see that perhaps this analysis reflects the *contested nature* of play within this game-based affinity space, in which the presence of the game's developer interacted with and occasionally clashed with the fan-based forms of play. The active participation of Ghostcrawler within the *World of Warcraft* case highlights an important finding in the evaluation of activities within affinity spaces: these forms of "ludic play" present within gaming affinity spaces can be seen as highlighting *multiple* and conflicting community framings of the activity within them. For the dedicated player such as Nawaf (or, for that matter, Namorada), the perspective of the community is obviously central, and the play of participants within the affinity space can be seen as a collective investigatory exercise, to understand the ways that the game *works* as in the sciences. For the game designer, Ghostcrawler, these activities can be seen as motivating exercises to drive player interest in the game, but they are purely academic concerns (in a pejorative sense), as they can lead participants within the affinity space to assume that they have more of a role in the game's design choices than they actually do. But, I argue, it is this tension that best characterizes the motivating dynamic of play within these affinity spaces—as Nawaf and Namorada both imply, they would love to see community science-like practices employed toward the design of the game, while Ghostcrawler is also very open about the role that scientific practice has played in his own path to becoming a game designer. The differences in how participants frame their conceptions of what the affinity space is *for* provided fodder for the further discussion, further argument, and further engagement with the game's systems in order to argue for the validity of one perspective over another.

This case may be the harbinger for similar conflicts in the future. With the persistent nature of online gaming fan communities and the increased desire of game companies to "tap into" the forms of motivating, engaging play that characterize them, I suspect that future gaming affinity spaces will feature similar issues. At the very least, we can see that Gee's notion of the affinity space needs to be further developed to accommodate the multiplicity of goals and commercial concerns that characterize the productive communities around major game franchises such as *World of Warcraft*. There are certainly multiple points of entry into the affinity space around a game, and multiple forms of play that participants may enact within them, but there are limits of expertise—and social structures—which constrain such engagement. The affinity space around a game such as *World of Warcraft* is a space of conflict and struggle for many of its participants (player/fan and professional game designer alike), and identifying the commitments that drive each position is a productive first step to unpacking how and why these spaces work.

In future research, I suggest that developing a better understanding of the commitments of individual actors within these spaces (as well as their larger personal and economic motivations) would seem to be a necessary next step. Developing affinity for a game and a community depends on a number of factors, and certainly not least among them is the issue of *power*, which is not thoroughly addressed in many analyses of these spaces. Though there are tantalizing learning and literacy practices within the online discussion forums around games such as *World of Warcraft*, we cannot fully understand why participants are doing what they do without acknowledging that forms of "play" can, in some cases, be instruments toward activities that are much more constrained and structured than the participants may realize. As with the sciences that Nawaf and Namorada wished to connect their affinity space play practices to, how and why participants engage in particular forms of research shape them toward desired ends. Affinity spaces may appear to adopt a life of their own, but are intentionally shaped by actors with commercial and professional concerns. In future analyses of gaming affinity spaces, I argue that we need to develop better ways of understanding the ways that participants' framing of their own activities clash with or are shaped by the desires of the media producers who choose to engage with fans in these spaces. A focus on the discursive moves presented by participants in an affinity space is a first step, but as research into this field matures, we need to better address the commercial and economic concerns that guide multiple actors' activities within the spaces.

NOTES

1. Talmadge Wright, Eric Boria, and Paul Briedenbach, "Creative Player Actions in FPS Online Video Games: Playing *Counter-Strike*," *Game Studies 2*, 2(December 2002), accessed at http://www.gamestudies.org/0202/wright/.

2. Johan Huizinga, *Homo Ludens: A Study of the Play-Element in Culture* (Boston, MA: Beacon Press, 1955).

3. Mia Consalvo, "There Is No Magic Circle," *Games and Culture 4*, 4 (October, 2009), 408–417.

4. James Paul Gee, *Situated Language and Learning: A Critique of Traditional Schooling* (London: Routledge, 2004).

5. See Sean C. Duncan, "Gamers as Designers: A Framework for Investigating Design in Gaming Affinity Spaces," *E-Learning and Digital Media 7*, 1(2010): 21–34; Rebecca Black, *Adolescents and Online Fan Fiction* (New York: Peter Lang Publishing, 2008).

6. See Gee, *Situated Language and Learning*, 85–89.

7. Jean Lave and Etienne Wenger, *Situated Learning: Legitimate Peripheral Participation* (Cambridge: Cambridge University Press, 1991).

8. Kurt Squire and Levi Giovanetto, "The Higher Education of Gaming," *E-Learning and Digital Media 5*, 1(2008): 2–28.

9. James Paul Gee and Elisabeth Hayes, *Women and Gaming: The Sims and 21st Century Learning* (New York: Palgrave Macmillan, 2010).

10. James Paul Gee, *An Introduction to Discourse Analysis, 2nd edition* (London: Routledge, 2006).

11. See MMOGChart.com, "MMOG Active Subscriptions," accessed at http://www.mmogchart.com/charts/.

12. Constance Steinkuehler and Sean C. Duncan, "Scientific Habits of Mind In Virtual Worlds," *Journal of Science Education and Technology 17*, 6(2008): 530–543.

13. Adam Holisky, "Ghostcrawler to Take a Break from the Forums" (WoW.com, 2011), accessed at http://www.wow.com/2009/09/26/ghostcrawler-to-take-a-break-from-the-forums/.

14. SimulationCraft, "SimulationCraft: WoW Raid DPS/TPS Simulator" (Google, 2011), accessed at http://code.google.com/p/simulationcraft/.

15. Toskk, "Toskk's Feral Spreadsheet" (Creative Commons, n.d.), accessed at http://druid.wikispaces.com/ToskksDPSGearMethod.

16. See MMO-Champion Blue Tracker (Curse.com, n.d.), accessed at http://blue.mmo-champion.com/.

Chapter Eleven

Cosmo-Play

Japanese Videogames and Western Gamers

Mia Consalvo

The first Japanese videogame that Omnistrife[1] remembers playing is *Final Fantasy VII*, when she was about nine years old. She recounted to me that at the time she "didn't know it was Japanese" and her introduction to it was rather accidental: her father had installed it on a computer in the house and said to her, "Okay, play with it until it breaks." And she did, getting about midway through the game before the video card died. Yet that early exposure resulted in an introduction to what would become her favorite game of all time; and along with that, a developing interest in Japanese popular culture and Japanese culture more generally.

While most Western player origin stories with videogames don't involve breaking a computer, many from the 1990s do involve playing Japanese games, on Japanese systems. Many players of those games were like Omnistrife, not even realizing that the "really good games" they were playing came from another country, and in particular one not Western. Yet because of the peculiar structure of the videogame industry at that time,[2] games did have a global distribution, and many (if not most) of the most-played games were created in Japan.

Another player I talked with, Ohako, observed that many of those early games had what he termed a "cultureless cover;" in that the "Japaneseness" of the games (or systems) was not obvious to the casual player. Some of that may have been a result of the technical limitations on games of that time period—the level of complexity in terms of graphics and dialogue was relatively simple, leading to fewer required changes than something on the order of a television series or feature film. And as Iwabuchi would likely argue, early Japanese game companies may have worked to eliminate any residual sense of foreignness that their games contained, through efforts such as local-ization and the adjusting of gameplay difficulty levels, in order to gain that wider market.[3] And so, Japanese games may have had easier entrée into foreign markets based on that simplicity, along with the marketing and busi-ness acumen of companies such as Nintendo or Square.

But what resulted was much more than success in global sales—it nudged some of those early game players into considering a world beyond their own, not simply in terms of the game space itself, but of the culture and country that had produced them. For some players, a global consciousness began to form, and perhaps even a form of cosmopolitanism emerged in their attitudes and world views. This chapter explores that process, and argues that Western players of Japanese videogames inhabit multiple spaces along a continuum of cosmopolitan dispositions. Further, it contributes to the discourse of cosmo-politan theorization by offering empirical evidence from an overlooked field—game studies.

COSMOPOLITANISM: HISTORY OF A TERM

Seeking to provide an over-arching definition that draws from much scholar-ship in this area, Skrbis et al. write "cosmopolitanism—as a subjective out-look, attitude, or practice—is associated with a conscious openness to the world and to cultural differences."[4] But beyond such a baseline, scholars disagree on how to apply or even understand the term. The history of the term(s) cosmopolitan/cosmopolitanism is long and complex, although the terms have recently become popular as theorists seek to understand how individuals as well as groups grapple with the processes of globalization that shape our world. In doing so, they have sought a concept that focuses on the human, and the role of the nation-state in a world that now seeks to alternate-ly dissolve as well as re-instantiate its various borders almost continuously.

In this debate, cosmopolitanism has been conceptualized both as a social *category* and as a social *ideal*, and as both a quality of individuals and as a general condition of a society or culture. Many definitions have been debat-ed, although the idea of the "world citizen" is always central in some way. Debates also focus on the centrality of mobility—is cosmopolitanism only present in those who have physically moved around the globe? It could be either for work (elite business travelers), lifestyle or work choice (expatri-ates), or perhaps even involuntary movement (such as refugees).[5] Further, scholars question whether individuals must *choose voluntarily* that they are open to other cultures and ways of life, or if this might be something uncon-sciously consumed, as we are all increasingly exposed to media, financial concerns, and ideological actions from diverse groups around the world.

More specifically, of what use is cosmopolitanism in thinking through the activities or identities of Western players of Japanese videogames? Most of the discussion of cosmopolitanism is abstract, taking little notice of actual daily practices by individuals or groups. But as some argue, "our understanding of cosmopolitanism should not be constructed from a series of imaginary, utopian, or ideal types; the fluidity and complexity of cosmopolitanism are only likely to be revealed by the empirical study of its mundane reality."[6]

A focus on empirical evidence is, however, somewhat rare in the discourse surrounding cosmopolitanism. A few scholars have tried to focus on particular elements of cosmopolitanism, arguing for example that we are seeing the development of "consumer cosmopolitanism," which is an openness to buying foreign brands, and possibly the welcoming of those brands, based on elements of exoticism and difference.[7] Likewise, scholars have argued in a similar vein that the spread of global media has given rise to either "pop cosmopolitanism" or "aesthetic cosmopolitanism," which focus not on buying products, but on consuming media that either comes from outside one's own (home) culture,[8] or through the gradual evolution of one's home media to take into account foreign influences, thus hybridizing the media of at least two cultures.[9]

Media scholars who have empirically investigated cosmopolitan attitudes among contemporary citizens have primarily focused on the global flow of media and its potential effects on creating global consciousness in general viewers. Urry is one of the few who has examined contemporary media for evidence of cosmopolitan images, finding that television offers us "globes, symbols, individuals, environments, trademarks and advertisements, that articulate a banal globalism, which point beyond national boundaries to the edges of the globe."[10] From those findings and in talking with media consumers, he concludes that "television and travel, the mobile and the modem, seem to be producing a global village," yet there are still few "global citizens" to be found.[11] Instead, most individuals "demonstrated a mundane cosmopolitanism within their daily lives."[12] Skrbis and Woodward further argue that individuals "reservedly deploy their cosmopolitanism" based on past experiences that can be either enjoyable or challenging.[13] And at perhaps the most pessimistic end of the spectrum, Halsall fears that "the barrage of global signifiers in the CNN broadcast" can lead to forces opposed to cosmopolitanism, which ultimately serve as "an immunizing and interiorizing force which might cancel out and indeed reverse any such trend."[14]

Despite such attention, there has been little application of cosmopolitanism as a conceptual category in the field of new media studies, or digital games studies more particularly. Part of that has perhaps been related to the difficulties of studying transnational fandom, but even when studying "local" fans, the tastes they have for global media such as anime, manga, or Japanese video games is not conceptualized through this lens. Henry Jenkins briefly addressed this topic and attempted to coin the term 'pop cosmopolitanism' to describe fans of global media such as anime, and Constance Steinkuehler has likewise invoked that concept to argue for the importance of studying virtual world spaces like Massively Multiplayer Online Games (MMOGs) due to their global player base.[15] However, these examples offer little detailed study, and no real exploration of the literature surrounding the study of cosmopolitanism, other than the surface invocation of these terms.

What then, would it mean if we described Western players of Japanese videogames as cosmopolitans? Would they all fit under such a label, or are there distinctions to be made between different individuals and their under-standings of, use of, and resulting beliefs based on their media use? And reflecting back, how does this category of players shape or alter theories of cosmopolitanism? Is movement between nations or cultures a necessary con-dition to achieving cosmopolitanism? Should multilingualism be similarly privileged? If not, if cosmopolitanism can be linked with openness to ideas, a willingness to learn more about other groups and societies, what does that mean for future society? Is this something to strive for, or instead a key trend that is important to map, yet lacking in any larger transformative value? This chapter will explore those questions through an analysis of players (if not fans) of Japanese videogames.

Setting the stage, most of the players I talked with for this project have used videogames as well as other popular Japanese media as a springboard for further exploration of Japan, its language, and people. Yet they them-selves do not represent all players of Japanese videogames, and I did not seek that wide a spectrum for this study. However, it might be useful to think of their activities (especially in relation to other Western players) as falling along a spectrum of cosmopolitanisms, particularly as their interests and actions can (and do) shift over time.

To situate this study, one end of the spectrum might represent consump-tion of Japanese media at its most basic and superficial—the "banal cosmo-politanism" mentioned by theorists such as Urry.[16] While I have consciously sought out individuals with interests beyond unreflective consumption, many of those same individuals admit that their early engagements with Japanese games began that way—many often did not even realize that the games they were playing (or anime they were watching) were foreign in any way. In-deed, some of the games or cartoons they watched were so localized that it would be difficult to "see" any Japaneseness within them, unless one was intentionally looking.

While terms such as banal, superficial, and unreflective may appear loaded, such a form of cosmopolitanism is not necessarily negative, or reflec-tive of unenlightened consumption. For most of us, our daily lives are full of much banal cosmopolitanism, as we buy strawberries from Mexico at the supermarket, t-shirts made in Bangladesh at the mall, and listen to BBC world news at home on the radio, after perhaps watching NBC broadcast the 2010 winter Olympics from Canada. It's difficult to believe we have the interest, let alone time and energy, to carefully investigate or pursue each strand of the global that we allow into our lives. However, that we do allow and even welcome it at all, even at the level of consumption alone, is indicative of a level of openness that indicates our willingness to see the foreign as more than an object of suspicion, and foreign peoples as perhaps not so different from us—or at least some of them.

Thus banal cosmopolitanism is perhaps a default position that most individuals in the West start from today, as we are inescapably surrounded by global media. Yet even at this level of consumption, for those who do not wish to pursue interests in another culture, language, or society further, there can be a particular interest or pursuit of media products that come from a specific tradition or culture, such as the stories and visual styles that encompass many Japanese videogames.

Players, Gamers, Fans?

I began this project with an interest in talking with individuals who expressed great interest in Japanese videogames, and who had gone on to investigate other areas of Japanese life and culture. In the process I have struggled with how to name these individuals. I do not believe they are all necessarily *fans* of Japanese videogames—indeed many of them expressed preferences only for particular genres (such as RPGS or fighting games) and not other genres with strong Japanese roots. And while their interests did extend beyond games, I hesitate to apply the label of "fan of Japanese culture" to them as well—that moniker is too sweeping, and seems somewhat dismissive of their actual beliefs and practices. So they remain "players" and sometimes "gamers."[17] But if individuals do self-identify as fans or gamers in a particular context that is important, I make sure to highlight that discourse.

Another group that has received much similar attention would be (self-identified) fans of anime and manga, studied extensively by scholars such as Napier and Levi.[18] Such research has investigated what draws Western fans to those particular media choices, what they find especially valuable and interesting about them, and fans' particular viewing and reading choices. A central finding is that such fans enjoy manga and anime for the stories they tell, which many fans believe are different from what they encounter in other popular, commercial media. As Napier explains, such fans believe that the stories found in anime are more compelling, more complex, and often conclude in ways different from Western stories. More specifically, Napier writes about fans who argue that anime tackles subjects deemed too sophisticated for younger viewers, and does not shy away from endings including death or defeat. Whether such representations of anime are true or not, that is how fans characterize them, and thus is important in understanding what brings them to anime.

Scholars have also looked at how fans have sought to expand their consumption of Japanese popular culture, through what Ito terms the 'media mix' that is offered by the industry.[19] This can include tie-in merchandise of the mediated and non-mediated sort, and various versions of a media brand, as it morphs across print, video, film, and even games. Likewise, scholars are also quick to point out how interest by anime and manga fans can lead to a larger interest in Japanese culture or popular culture, and in particular how the act of studying the Japanese language has been popularized, thanks in large part to the spread of anime in the West. Yet although scholars take a critical eye to this media mix and acknowledge its component parts, videogames are usually given short shrift. If mentioned at all, it is usually in reference to a mega-franchise such as Pokémon, and the specificity of the various games involved is not fully examined. In contrast, this chapter aims to take games as a starting point, and explore how Japanese videogames are important in the lives of game players, and how they too can make a contribution to a players' sense of themselves, and their interests in other cultures and in the global condition of cosmopolitanism.[20]

ORIGIN STORIES

For most players, their entry point was indeed some form of Japanese popular culture, but for a minority of individuals, interest arose from childhood experiences that emphasized cultural exploration, or an individual person who sparked a desire to learn more about a particular country and way of life that seemed very different. For example, Kelly describes going to the library as a child and reading "all the children's books on Japan or things related to Japan that I could find, which meant I ended up reading about things ranging from food to World War II internment camps in the US." That led to her eventual interest in the popular culture of Japan, including anime and videogames. But for most individuals, popular culture was the entry point, and they were introduced either by a friend or family member. A typical story comes from Omnistrife, who describes her interest starting with videogames around the age of 9, and *Final Fantasy VII* for the PC in particular: "my dad bought the game for himself, and then he kinda played it for a while and I guess he lost interest and, I had a friend who ranted and raved about it so I popped it into the computer and I've been playing it ever since. It's my favorite game, in the world." Similarly, Doris was first introduced to Japanese videogames through her dad and brother, who were playing *Final Fantasy VI* "and that led me to play the game, play more RPGs like it and then from then on . . . I started getting into it."

Others were first exposed to Japanese popular culture through anime, by friends—either through a specific series such as *Sailor Moon*—or through finding blocks of anime on television like the cable channel SciFi's Japanimation lineup and Cartoon Network's Toonami series. For Jean, that interest was gradual—starting with the *Sailor Moon* series—but then "eventually I got hooked and [even though] I didn't like it—I didn't even *know* it was from Japan actually, and I was just watching it 'cause it was on TV. But, the show stopped at a certain point on Toonami, when it was on Cartoon Network, and I really wanted to know how it ended and so I found somebody online who was like 'I have tapes! VHS tapes, of the Japanese with English subtitles and I'll sell them to you.'" Jean went on to find other anime series from Japan she thought she would like, and also began searching for manga that might also be to her liking. Such snowballing of interests is common, especially when there are tie-ins across a particular content universe, such as the *Ranma ½* or *Bleach* series that can span anime, manga, and games, making consumption more obvious for fans of a particular series. However, not all individuals find different components of a media mix equally compelling—Omnistrife for example saw herself as a fan of the *Naruto* anime series, yet did not like the related games, as they were fighting games. She clarified that "if the anime creators could come up with a videogame that actually goes along with the story of the anime, you know, it could have a plot. Then I would definitely like to play them." Thus, genre did matter to individuals, and different pieces of a media franchise or mix could meet with different levels of success, based on whether they drew from the same elements across media that particular consumers enjoyed.

Age factored differently into when individuals started playing Japanese videogames, or when their interests in Japanese culture began. Like Kelly, Miranda reported being interested in Japanese culture from a young age, when she would receive packages of educational materials about different countries in the mail via subscription. Her interests in Japan's popular culture followed suit, as she played games and watched anime at the same time, and at some point made the connection that many, if not most, of the games she enjoyed playing were from Japan. Others developed their interests later on, often in high school or even later, through friends with those interests. Anna, for example, describes herself as a "late bloomer" in terms of playing videogames—her first real experience was playing *Kingdom Hearts* at the age of eighteen. After that, however, she ended up buying a Playstation 2 and her own copy of the game, and then started exploring other titles.

In terms of the titles or genres that these players enjoyed, most started out playing games considered very popular or mainstream for their time, including various titles from the aforementioned *Final Fantasy* series, *Super Mario Brothers*, and other games such as *Pokémon* and the different *Zelda* titles. In contrast, first experiences with anime (for most individuals I talked with, anime viewing came before they started reading manga) could draw from either popular commercial products officially released in North America such as *Sailor Moon*, or it might center on more obscure titles found by friends, such as *Ranma ½*. That likely reflected the different distribution and technological constraints of the two mediums of anime (videotape) and videogames (cartridges and CDs-ROMs).

At that time (1990s) there were few anime productions (either films or television series) commercially available in the West, although the format for viewing—VHS—made it relatively easy to acquire or copy offerings not already broadcast, thus easily increasing the potential pool of offerings. In contrast, the videogame industry was already building a strong pipeline from Japan to the West, with many (if not most) best-selling games at that time coming from Japanese developers. But if Western players wanted to play games not (yet) sold in the West, they would be required to modify their consoles—a much more complicated process than seeking out VHS tapes. Thus anime viewers were more able and thus likely to see a wider selection of materials, although game players did have a good range of titles available for them to play.

Given what we know of fans of anime, what draws individuals to Japanese videogames, particularly those who acknowledge and appreciate their Japanese origins? Some of the same themes emerge, although there are some differences, due to the qualitative difference—namely interactivity—between games and other media. For many players, though, story is a central component driving their interests. As Omnistrife explains, her expectation is that Japanese games simply have superior stories, "I just kind of expect stories to be better than with American games. . . . Japan has this like higher status in my mind. It's just like they have more creative people who can come up with better stories." In explaining what she means, she elaborates that "the characters fascinate me, the loopholes in the story fascinate me, I love trying to fill in the missing information . . . like . . . how long was Vincent Valentine [a character in *Final Fantasy 7*] really sleeping in his coffin and who is Sephiroth's real father."

Other players also have high expectations for stories in Japanese games, as well as the characters and other elements such as visual or musical style, even if they do not actually enjoy playing games with deep stories. Jean relates that she actually prefers games that are short, often comical, or fun. She does not play RPGs, even though she acknowledges, "I love the stories," but at the same time, "the gameplay, I guess, bores me." That sense of admiration for story does carry into the games she actually plays, such as the *Silent Hill* series of horror games. She explains that she finds it fascinating that in the second game of the series "you find out that he [James Sunderland, the main character] killed her [Mary, his wife, whose death he is investigating]; the whole thing is just, whether he's tried to block it from his memory or, it's like the town pulling that out of him and making him live the horror that he visited upon her."

While Western games can also have good stories, which some players did acknowledge, certain elements of Japanese games seem to qualify as different or notable in some way. Kelly talks of the "interesting worlds and people who do amazing things" that comprise many stories in Japanese games. Such components can serve as more of a backdrop for something like fighting or music/rhythm games, or the central component to a role-playing game. Moti explains that with the Nintendo DS game *The World Ends With You*, for example, "I did like the story, I thought it was engrossing and it was fun . . . the moral of the world itself got unlocked. . . . It felt more like the real world possibly because it was really just sort of a little variation on the real world." Similarly for Doris, *Final Fantasy VI* is different from other games where "the princess always gets the prince at the end of the story"; and where instead "it wasn't lying to me about life or anything like that. . . . It wasn't a Hollywood ending. My favorite character dies at the end, and . . . . that was beautiful to me." For Ohako, the difference can lie in how Japanese myths and legends find their way into games, such as the *Phoenix Wright* series of law/adventure games where even though the localizers placed the story in America, still "the game is full of Shinto priests and possession and magatamas and all kinds of that imagery."

Another notable element that made Japanese games different or special was their particular sense of style, which varied based on the genre or type of game being discussed. Two games mentioned by various players as particularly Japanese in their style and execution were *Okami* and *Katamary Damacy*. *Okami* is the story of the Japanese sun goddess Amaterasu, appearing in white wolf form, who returns to the world in order to save it from darkness. The game was hailed for its art style, which draws from sumi-e (ink brush) and cel-shading techniques for its particular look. It also features as a central game mechanic, drawing with a calligraphy brush, which changes the world in various ways. That mechanic was remarked on by players as being distinctive, innovative, and (obviously) very Japanese in its design. Thus gameplay, itself, could be expressive of Japanese culture, albeit normally not in such an obvious manner.

Katamari Damacy was developed by the artist Keita Takahashi, and centers on the son of the King of All Cosmos, who must "roll up" collections of various items to form stars and send them into the sky. The game is heavily stylized and non-realistic, featuring people, cows, buildings, and more as material suitable for the construction of katamaris. Interestingly, Jones the-orizes that the central activity of this game—collecting things—is a commen-tary on the somewhat obsessive collecting activities of otaku in Japan (and likely elsewhere), thus drawing from Japanese culture again for a central gameplay mechanic. [21]

Overall, players felt that such games were special, and that style—partic-ularly visual style—was a central element driving their interest in playing such games. Indeed, Ohako repeatedly referred to "style" when asked to discuss his favorite games, and why they were chosen. Each game, he ex-plained, had a unique visual style, whether it related to 2D sprite graphics for a fighter game, or the wood block print style employed in *Okami.*

Most players easily called out unique graphical elements, sometimes also drawing in the musical scores or soundtracks to their favorite games as something that was different from other games they had played. Ohako noted that the game *Visions of Mana* had an "arresting" visual style, with "colors everywhere" and "detail in everything." Likewise Miranda jokingly recalled how the design of the creatures in *Patapon* were "so freaking adorable" at the same time they were "these killing machines."

Doris similarly recalled how the artwork in *Final Fantasy VI* was compel-ling to her, with a style by artist Yoshitaka Amano that was "very very free flowing, very wispy. And the characters. . . . His style is very similar across the way, he's very much into like little details and a lot of um . . . he puts a lot of pretty things in there, like, they're wearing scarves and tassels and there's all these little symbols all over them and it's all very very pretty."

Overall then, players believed that style was a central element in their enjoyment of Japanese games. It was something they found unique or special in particular games, although there was no unifying stylistic theme they associated with Japanese games in general, except perhaps for something notably absent—graphical realism—but quite prevalent in Western games. So aside from being colorful (whether in bright or muted tones), style usually meant visual imagery that departed in some way from typical (Western) videogame conventions, which they found compelling or simply satisfying.

THE JAPANESENESS OF STORY; DIFFICULTIES OF GENERALIZATIONS

I was particularly interested in how players saw Japanese games in relation to their origins—i.e., their Japaneseness. Discussion of this element of games (as well as some anime and manga) proved a bit difficult for some respon-dents. Many respondents were unwilling to assign broader cultural meanings to the themes that they found in videogames, but many could indeed identify themes or approaches that seemed distinctly different from those in other (Western) videogames. For some players, their early game choices were really about playing "good" games, or games recommended to them by friends and family, and it was only through gameplay that they discovered those games were Japanese, or that a common thread linking their interest in particular games was their Japaneseness. Of course that wasn't the case for all players, and many do play and enjoy Western games. But it should be noted how most early gameplay experiences were situated as games generically, although as Ohako explains, the days of "the cultureless cover of Japanese game development [are] gone for good," and we can expect to see more games both with highly sophisticated localization, as well as some with "a more distinctly Japanese flavor."

For some players, cultural expressions relative to Japan could be found in a game's inclusion of utilitarian knowledge about Japanese daily life or culture. For example, several players who had played *The World Ends With You* noted how the game drew on actual life in Tokyo and its various neighborhoods and cultures, which helped them to learn more about the geographic layout of the city. Miranda remarked that playing the game actually helped her "get around Shibuya when I lived there." Likewise, more than one player mentioned the *Persona* games as offering insight into daily life for teenagers in school in Japan. So while the plot of the game involves high school students who must kill demons, players remarked that the game also taught them about the school system in Japan and the importance of exams, and how those structures differed in key ways from Western systems. Similarly, while playing *Pokémon* Miranda always found it challenging that the main store where players could buy particular items had different levels, and your avatar had to go to different levels to purchase different items. Upon visiting Japan, she learned that "if you go to a department store in Japan, it's the same thing." Thus, even seemingly minor gameplay elements such as the geographic location or mechanics such as stores can offer players keys to thinking about cultural differences.

Players also noted how characterization in Japanese games could be different, particularly the ways in which characters might express certain worldviews. Miranda related how she found values being expressed differently in Japanese videogames, such as in *Pokémon* when "it always talks about how you have to believe in your heart and . . . if you talked about them in American games, it'd be like 'what are you talking about? Like, no, you didn't win because you believe in your Pokémon more than the other person. No, you won because you were stronger,' you know, that's the American take on that." Likewise, Omnistrife relates how she felt the lifestream depicted in *Final Fantasy VII* "makes you think that [it's] like their religion.

And I think that a Western culture which is traditionally Christian or sometimes pagan or whatever, I don't think they would think of something like that."

Some players identified larger themes that appeared Japanese in origin. Ohako explained he found a "strong sense of hope and fantasy" in Japanese popular culture (inclusive of games, anime, and manga). He elaborated "in the US, it feels to me like the main lynchpin for drama is fear (hostage situations, serial killers, evil mothers, etc.), while the main drive for drama in IJPC [Imported Japanese Popular Culture] is hope (wielding pendants/robots/ hot cyborg ladies/large swords to win against the bad guys)." Talking of the contradictions involved in Japanese culture, Moti argued that "the more popular figures in Japanese history are often the people who lost. You know, they lost nobly." Kelly believed that Japanese games can also draw from more universal themes about the hero's journey, conflict between countries, and loyalty to one's family and hometown. Yet she also argues that Japanese games "tend to emphasize being part of a team or a group a bit more than most Western games do." As an example, she explained the differences between playing *World of Warcraft* (a US-developed MMOG, also referred to as *WoW*) and *Final Fantasy XI Online* (a Japanese-developed MMOG, also referred to as *FFXI*) and how *FFXI* is a much more group-oriented game than *WoW*.

Doris went beyond the level of games, to argue for how different console systems might reflect different cultures or values as well. For her, the Japanese Wii is "very private and you don't really get to talk to a lot of your friends unless you know their code, you have *their* code, there are codes to get there. And you pretty much need to know the person *in person* in order to play online with them." In contrast, the more Western Xbox assumes "you're pretty much using the Internet to play with other players. So I think that that kind of speaks for the type of videogames that they're interested in. Like, [the Wii and Japanese culture is] less about really playing with strangers and more about playing by yourself," or only with those who you know.

In addition to seeing themes or values that might be expressed differently, or stories or characters they felt were unique, players were often unapologetically interested in and took pleasure in playing a game that originated in a culture not their own. Anna summed it up, stating "I really enjoy games when they are unabashedly Japanese." Similarly, Omnistrife felt that "*Persona*'s a very *very* Japanese game" and that was part of what she enjoyed about it. Miranda believed that many Japanese games embodied a certain silliness, a sense of not taking themselves too seriously, even as they tried to impart a message at the end, which was something she enjoyed about them. Other players echoed that sentiment, seeing certain Japanese games as "goofy" or "comical," yet in ways that made them endearing.

Such beliefs mirror or build on Napier's findings concerning Western fans of Japanese anime—that they recognize something different, something foreign in their object of interest, and that difference is often the element that is so compelling.[22] So rather than shed the "cultural odor" Iwabuchi argues Japanese products have traditionally tried to eliminate in their global travels,[23] Japanese games often end up keeping many of those elements, much as Allison describes other contemporary toys doing, as Japanese popular culture continues to be popular outside its home region.[24]

For some, an interest in and enjoyment of the stories offered by Japanese games such as *Silent Hill, The World Ends With You*, and *Persona 3* and *4* is an end in itself. These players enjoy the complexity of the characters and mature storylines that don't always promise happy endings. Players also point to the unique or special *styles* invoked by Japanese games, both visually and aurally. Players mention the soundtracks of various games as a particular draw, and most all players I talked with expressed great appreciation of the different aesthetic styles offered by various Japanese games. Thus, the wood-block print style of *Okami*, the fanciful, delicate aesthetic of an artist for *Final Fantasy*, and the flashy colors of *Kingdom Hearts* all drew individuals to those games, and captivated them. And for most Western players generally, this is the terminus of their interest in Japan or its culture.

But for the players I talked with, their interests instead widened at this point, as they sought out more games, more popular culture artifacts, and then more exposure to Japan as a society, with a distinct language, culture, and way of life. These individuals have moved beyond banal cosmopolitanism, embracing a deeper interest in Japanese culture. Players began to express interest in games specifically due to their Japanese origins, or were often interested in finding Japanese games not released in the West. That might lead for some to language study to access those products more successfully. And for all at this point, there is a discourse of virtual travel at play—as players seek the mobility of exploration through games, through virtual exposure to another culture. For others, there is the mobility not simply of virtual travel via media, but physical travel, as many players have spent time in Japan, either through formal study abroad programs via their universities and colleges, or through their own independent travel.

IMPORT GAMES, MODS, AND MODDING

The virtual travel that some players found enjoyable upon initial play of Japanese videogames sometimes deepened into an interest in finding Japanese games that either had not yet been (or would never be) released outside Japan, or to games they might have already played, but wanted to experience in Japanese. However, to play such games, some ingenuity was necessary. Most game consoles are region-encoded, meaning that when games are released, they are designed to run only on systems in a particular region— Japan, North America, Europe, or International. A few hardware systems, such as Nintendo's Game Boy and DS series, have historically been region-free, but that is the exception rather than the rule. For players wishing to play Japanese region-locked games on other systems, they would need to either modify their own console (which usually meant having some technical skill and voiding the warranty), or purchasing a Japanese version of that same console.[25]

Not many players I talked with went this far, with most expressing satisfaction with the games they could play on their own systems, in English. A few had considered their options and decided against doing so, either because of the costs involved in buying another system, or fear of "messing up" the hardware they or a friend currently owned. A few more tried intermediate techniques, such as Kelly, who bought an adaptor for her SNES in order to play an imported copy of *Final Fantasy V*, which had not then been released in the United States. Similarly Doris used an R4 cartridge with her Nintendo DS to play "home brew" games. Ohako bought Japanese language games for his Nintendo DS, such as the *Ouendan* music game, because it came with a better selection of songs than the North American version.

Similar to console game players generally, only a minority of those I talked with admitted to modding their console, or buying a modded console in order to play particular games. A few did actually purchase Japanese consoles. Dan bought a PS2 when he was in Japan in 2003 in order to play some games, and Moti was the most active, as he had modded a Super Nintendo system, bought a modded Playstation, and more recently purchased a Japanese PS2. He did so in order to play games not available outside Japan, such as the *King of Fighters 2002 Unlimited Match* and the *Nana* game based on the manga of the same name. Such gameplay in Japanese also helped him keep his language skills up. Yet for this group overall, that level of activity was the exception rather than the rule.

While it is difficult to say how widespread the practice is of playing Japanese-language games among Western players, the individuals I spoke with often mentioned the challenge of doing so, whether they actually played them or not. Even for those studying Japanese (discussed next), text heavy games presented a huge barrier to entry for most players, with their reliance on the highly complex kanji system of writing. Only the most dedicated (and fluent) gave more than a passing attempt at doing so, either sticking to games that had easily accessible commands and rules or playing only localized games. While there are communities of Western players who do translate such games for a wider audience, the majority of those I talked with seemed satisfied with the games they found—either localized versions, Japanese versions playable on their North American consoles, or occasionally Japanese language games they found particularly compelling. As with any level of playing or fandom, as practices become more specialized and demand more investment from the individual in terms of skill or expertise required, the number of individuals willing or able to engage in those practices understandably shrinks. So here the willingness or eagerness to explore additional facets of Japanese culture was met with the very real limits of language fluency and technological barriers—challenges that some embraced but many more did not. Yet even though most Japanese-language games were too daunting for these players, that didn't stop them in their efforts to learn at least some Japanese, either formally or informally, from very beginning levels to full fluency.

LEARNING JAPANESE

With one exception, all the individuals I talked with were currently studying or had studied Japanese in the past, with different levels of dedication and formality. A few were learning on their own, such as Ohako (who had purchased the DS title *My Japanese Tutor* to learn) and Anna (via books on how to converse in Japanese, CDs, and most recently kanji and hiragana practice). Most had taken classes in college, ranging from one semester to four years of classes, including a study abroad period in Japan. Time spent in coursework reflected varying levels of dedication. Doris took two years of Japanese mainly for her interest in the culture and the writing system; likewise Anna's self-instruction was driven by her interest in Japanese popular media, as well as her interest in languages generally. Omnistrife, who was the only one not interested in learning the language, had still enrolled in a Japanese culture class in order to learn more about daily life in Japan.

For those who considered themselves particularly dedicated, goals included becoming proficient enough to achieve certain levels in the standardized Japanese Language Proficiency Test (JLPT); to study abroad and converse with Japanese individuals; to work in Japan; and to base a career on the interest. Moti, for example, had achieved top-level certification for his Japanese and at the time of our interview was a graduate student getting a degree in linguistics. His undergraduate major was Japanese and he had lived in Japan for more than two years. Similarly, Dan started learning Japanese via manga and anime, then formalized his study through college level courses, study abroad, and then two years living in Japan. Miranda had self-taught herself some hiragana in high school, and in college took formal classes and studied in Japan for six months. While her study of kanji has since "dropped off," she was applying for an internship program to work in Japan for a summer when I interviewed her.

For most individuals, their interest in studying Japanese was driven at least in part by their enjoyment of Japanese videogames and other media. Some specifically mentioned that linkage—either as a spur to learn more Japanese, or as a way to keep fluent or practice the language. It also helped some of them with further interests, such as the creation of stories, fan fiction or fan art related to Japanese media. Such study has been noted by others as a recent phenomenon, with a surge of interest in college (and high school) Japanese language classes by students who have initially encountered the language in their pop culture consumption and now wish to learn more about the culture and perhaps gain access to cultural products not yet (or ever to be) translated or localized for Western audiences. One professor of East Asian studies had a student take her Japanese history class because he had become fascinated with the subject after playing videogames such as the 2005 *Genji: Dawn of the Samurai*, which was loosely based on the life story of the 12th century Japanese general Minamoto Yoshitsune.[26] And for most of this group, an interest in the language led to (or conversely was inspired by) the desire to visit Japan and explore it in person.

VISITING JAPAN

Most players I talked with had either already been to Japan or planned to visit it in the future. About half had actually been there, either for a vacation trip or more usually as part of a study abroad or work program. Only one person—Omnistrife—had no plans to go. It was not based on her lack of interest in the culture, but rather her dislike of travel, and unease at being in a place where she could not speak the language. For most, however, their interest in Japanese videogames had extended to wanting to see the culture and geography of Japan. Those that went talked of visiting regular tourist venues such as Tokyo, Kyoto, and Nara, and visiting shrines, temples, and museums. Likewise, more pop culture oriented venues such as the Studio Ghibli Museum were visited, as were places off the beaten path, including the countryside and more normal, everyday areas in Japan.

For some, a central part of the experience was interacting with Japanese people, something that paradoxically could be difficult if one was part of a study abroad group. As Jean explained, "it was hard to get away from the other Americans because we all wanted to stick together," but slowly she pushed herself to spend time with Japanese people, and she noticed her

Japanese language skills improving as a result. But more important perhaps was the overall experience—Jean explained that as a white woman in Japan, "this was one of the best experiences of my life, being a minority."

While all players may not have had such a transformative experience, many were particularly interested in seeing what 'daily life' in Japan was like, especially outside of major cities, and they also investigated other aspects of the culture they found interesting, including kabuki, the food, seasonal festivals, music, and dramas. Moti in particular has spent significant time in Japan, both for study and work, and while there has taken classes in ikebana and calligraphy, and attended theater performances of noh and bunraku, listened to shamisen performances, and learned kendo.

One of the more interesting travel stories arose from Miranda's interest in the *Phoenix Wright* videogame series, which is very popular in Japan. Before visiting she learned that an all-women theater group in Tokyo had created a musical version of the game, and she was determined to see it. So she traveled by herself to Tokyo (from Nagoya) for the event, where she was "the only foreigner there" in a crowd that was also "85–90% female." Describing the show as alternately bizarre, weird, and hilarious, she later mentioned to me that she might provide a translation of the musical for her friends, as the show has just been released on DVD. Thus player interests in Japan span a range from the traditionally tourist to deeper investigations of daily life, as well as events and artifacts springing from the interest that initially spawned their visit—videogames.

CONCLUSION

One of the last players I talked with about his interest in Japanese videogames was from Europe, and he had decided to become a game programmer based on his love of videogames, and his introduction to Konami games in particular. Along the way Darius had done many things in pursuit of his passions—he had modded two consoles in order to play Japanese games, he had translated a book, subbed an anime movie, and most impressively, taught himself Japanese well enough to achieve the first level of JLPT proficiency, and continues to study the language, even without ever visiting Japan (although it is on his to-do list). Thus Darius is perhaps the perfect example of a cosmopolitan citizen, or someone demonstrating elements of cosmopolitanism as it is constituted in contemporary culture.[27]

But more broadly, what do we know about players of Japanese videogames after hearing from individuals such as Darius, Omnistrife, and Doris? Obviously this is a special group, not a randomly selected group of individuals or even game players. I specifically sought out players who were passionate about Japanese videogames, and who had, based on that interest, taken their hobby further by investigating other aspects of Japanese culture. Thus, they are not comparable to a focus group consisting of everyday citizens questioned on their views of global images or news, nor are they consumers being tested for their openness to buying foreign brands.[28] Yet even with their expressed interests on the table, there are differences that emerged in their perspectives, both across the group and individually, as they explained their interests over time.

As I mentioned earlier, we can see some elements of mundane or banal cosmopolitanism in the early playing experiences of many gamers, as they chose (or were given) simply "good games" to play. In part due to the structure of the videogame industry, they were open to a product they either did not realize was foreign, or perhaps did, but in either case they saw nothing remarkable in that fact, being more excited by an interesting story, novel graphics, or innovative gameplay mechanics. Yet for most, that interest grew, and they became more consciously aware of the origins of those games, and more carefully began to seek out similar games, and then looked to the culture that created them as something else to be investigated.

While they were indeed all exhibiting the "pop cosmopolitanism" advocated by Jenkins, I believe we need a more sophisticated way to explain how they expressed their interests, rather than the binary of consumption/non-consumption as signified by that term. That term as defined also restricts its meaning to consumption, a concept too limiting to describe what I found among these players. Here, Urry can help again, through his formulation of the concept of "cosmopolitan predisposition." He believes that someone with such a predisposition can display the following practices and/or beliefs:

> extensive mobility in which people have the right to "travel" corporeally, imaginatively and virtually; the capacity to consume many places and environments en route; a curiosity about many places, peoples and cultures and at least a rudimentary ability to locate such places and cultures historically, geographically and anthropologically; a willingness to take risks by virtue of encountering the "other;" an ability to map one's own society and its culture in terms of a historical and geographical knowledge, to have some ability to reflect upon and judge aesthetically between different natures, places and societies; semiotic skills to be able to interpret images of various others, to see what they are meant to represent and to know when they are ironic; an openness to other peoples and cultures and a willingess/ability to appreciate some elements of the language/culture of the "other."[29]

That list echoes the stories of players described here, as they talked of their varying experiences. For all, openness to other peoples and cultures resulted from their game playing, as they wanted to learn more about the culture behind the games. Most also took up the challenge of investigating the language of that culture, with greater to lesser degrees of fluency obtained. Many also traveled "virtually" to Japan or to an imaginary Japan through their gameplay—particularly players who were interested in games exhibiting a high degree of "Japaneseness." Thus, games like *The World Ends With You*, and those in the *Persona* series gave players a sense of insight into Japanese geography, school culture, and the daily life of teenagers in Japan. Others did more, taking risks by attempting to encounter the "other"—here Japanese individuals—either through study abroad, travel, or talking with Japanese exchange students at their home location. For many this involved more than sharing stories of favorite games or language practice, but instead went to investigating daily life practices and perceived or real differences between the cultures. And some players also investigated the history of Japan, learning more about its geography, its past, and societal structures.

Thus, the players I talked with did indeed exhibit many facets of a cosmopolitan predisposition, much of which was spurred from their initial game playing. And it went beyond consumption of foreign media, to active exploration, as well as construction of their own creations (which I will discuss in future research). Clearly not all game players behave this way, but this activity does point to how one form of media—videogames—can indeed serve as not simply a point of contact with "otherness" but also as a catalyst for further exploration. This catalyst was not accidental—the structure of the game industry in the 1990s relied heavily on Japanese games and systems, and thus it might be considered inevitable that this happened. But this deep exploration does contradict Halsall's assertion of an "immunization" to what is foreign. It instead points to Regev's suggestion that cultures do mix as they come into contact—as early Japanese games have influenced the designers of today, both in Japan and in the West. And some players will eagerly seize on those elements and innovations, transforming themselves in the process.

I have investigated what cosmopolitanism has to offer to a study of game players, but it is also important to determine what such a study says about current theorizations of cosmopolitanism. Clearly, the study indicates that there are different levels of adherence to the model or ideal, and a person's position can change over time. This is a study that emphasized positive exposure to foreign media, so it might be natural to see openness emerge. And the structural forces of this industry were unique in offering players such a wide range of Japanese games to play. Yet more than this, this study argues for the importance of conducting empirical work, and in digging deeply into pockets of media or culture that may not be considered mainstream. Such components may indicate a vanguard for coming changes, or they may simply point to key events to understand now, which may go no further than a particular target audience. But the drive to understand cosmopolitanism must take actual individuals into account, and perhaps, I would argue, it is better to start with individuals and explore their particular areas of interest and how they have developed in relation to "foreign" media, rather than begin with broadly generalized images of globes or news. Because it is in these interesting spaces, where Western players learn about how to "believe in your heart" in order to win, where some of the most interesting components of a cosmopolitan predisposition may be found.

NOTES

1. All names used are pseudonyms.
2. Most players I talked with started playing videogames sometime in the 1990s, when Japanese companies such as Nintendo, Square, and Konami produced a large number (if not the majority) of games released globally. They were unfamiliar (except as a historical event), with the Atari 2600 system, and thus associated their early gameplay experiences with Japanese products.
3. Koichi Iwabuchi, *Recentering Globalization: Popular Culture and Japanese Transnationalism* (Durham, NC: Duke University Press, 2002).
4. Zlatko Skrbis, Gavin Kendall, and Ian Woodward, "Locating Cosmopolitanism: Between Humanist Ideal and Grounded Social Category," *Theory, Culture & Society 21*, 6(2004): 115–136, 117.
5. Skrbis, Kendall, and Woodward, "Locating Cosmopolitanism."
6. Skrbis, Kendall, and Woodward, "Locating Cosmopolitanism," 121.
7. Petra Riefler and Adamantios Diamantopoulos, "Consumer Cosmopolitanism: Review and Replication of the CYMYC Scale," *International Marketing 62*, (2009): 407–419.
8. Henry Jenkins, *Fans, Bloggers and Gamers: Media Consumers in a Digital Age* (New York: NYU Press, 2006).
9. Motti Regev, "Cultural Uniqueness and Aesthetic Cosmopolitanism," *European Journal of Social Theory 10*, 1(2007): 123–138.
10. John Urry, "The Global Media and Cosmopolitanism," (Department of Sociology, Lancaster University, Lancaster, UK, 2003), 5, accessed at http://www.comp.lancs.ac.uk/sociology/papers/Urry-Global-Media.pdf.
11. Urry, "The Global Media and Cosmopolitanism," 12.
12. Urry, "The Global Media and Cosmopolitanism," 9.
13. Zlatko Skrbis and Ian Woodward, "The Ambivalence of Ordinary Cosmopolitanism: Investigating the Limits of Cosmopolitan Openness," *The Sociological Review 55*, 4(2007): 730–747, 745.
14. Robert Halsall, "Towards a Phenomenological Critique of 'Mediated Cosmopolitanism,'" (paper presented at the annual meeting of the International Communication Association, Dresden, Germany, June 2006), 15.
15. Jenkins, *Fans, Bloggers and Gamers*; Constance Steinkuehler, "Virtual Worlds, Learning & the New Pop Cosmopolitanism," *TCRecord: The Voice of Scholarship in Education* (2006), accessed at http://www.tcrecord.org/PrintContent.asp?ContentID=12843.
16. Urry, "The Global Media and Cosmopolitanism."
17. The term "gamers" is also somewhat loaded—some players I talked with would not describe themselves as hardcore gamers, even if they played for many hours a day at a time.
18. Susan Napier, *From Impressionism to Anime: Japan as Fantasy and Fan Cult in the Mind of the West* (New York: Palgrave Macmillan, 2007); Antonia Levi, "The Americanization of Anime and Manga: Negotiating Popular Culture." In *Cinema Anime: Critical Engagements with Japanese Animation*, ed. by Steven Brown (New York: Palgrave Macmillan, 2006): 43–63.
19. Mimi Ito, "Hypersociality, Otaku, and the Digital Media Mix" (paper presented at the annual Society for the Social Studies of Science Meeting, New Orleans, LA, November 2002).

20. This work is based on a dozen qualitative interviews I have conducted with a variety of individuals, all of whom enjoy playing Japanese videogames, and have greater to lesser interests in other forms of Japanese culture. It also draws from my experiences interacting with similar sorts of individuals who played *Final Fantasy XI*, *Phoenix Wright*, and from various groups found online. Just as with any group of gamers, this is a diverse group that enjoys playing different types of games, for a variety of reasons. Individuals have different levels of interest in Japanese culture, and they play various roles in contributing to the culture and community surrounding Japanese popular media.

21. Steven Jones, *The Meaning of Video Games* (New York: Routlege, 2008).

22. Napier, *From Impressionism to Anime.*

23. Iwabuchi, *Recentering Globalization.*

24. Anne Allison, *Millenial Monsters: Japanese Toys and the Global Imagination* (Berkeley: University of California Press, 2006).

25. Mia Consalvo, *Cheating: Gaining Advantage in Videogames* (Cambridge, MA: MIT Press, 2007).

26. Shelley Fenno Quinn, personal communication, October 16, 2009.

27. Urry, "The Global Media and Cosmopolitanism."

28. Urry, "The Global Media and Cosmopolitanism"; Riefler and Diamontopoulos, "Consumer Cosmopolitanism."

29. Urry, "The Global Media and Cosmopolitanism," 7–8.

Chapter Twelve

Beyond the Virtual Realm

Fallout *Fans and the Troublesome Issue of Ownership in Videogame Fandom*

R. M. Milner

LABOR AND LEISURE BEYOND THE VIRTUAL REALM: THE GAMER, THE FAN

My name is R. M. Milner, and I am a gamer. And moreover, I do not game alone. I air out the football in *Madden*[1] with my brother next to me on the couch. I pluck my *Rock Band*[2] bass with my back against the guitarist and a crowd cheering from behind the couch. I set up my television next to the TVs of three friends in a basement as we "party up" in *Gears of War*. We then jump into online deathmatch rooms and take on all comers, some of which are four or five strong in a basement across the country or world.[3] When I am not *playing* games, I am browsing game-related media on my Xbox. I am looking up tips and strategies on my iPhone. I am watching videos, reading reviews, and gawking at forums on sites like *IGN*, *Game Spot*, and *Game Trailers.*[4] I might even venture over to sites devoted to specific titles, genres, or eras in gaming (both "official" and "unofficial"). I will often post what I see on friends' Facebook profiles. I am not just R. M. Milner the gamer, I am R. M. Milner the videogame *fan*. And, increasingly, it is harder to separate one from the other. Far from the stereotype of the lonely soul bunkered in a windowless room, gaming is an inherently *social* medium,[5] a fan medium.

While gamers have long engaged in fan activity and have produced a complex and specialized subculture, the literature surrounding the two subjects has mostly been separate, with little crossover between gaming scholarship and fan scholarship. It has only been recently that the likes of Crawford

and Rutter, Jenkins, Newman, and Taylor[6] have argued that we begin looking at the social interaction that surrounds the game. For many gamers, the
time spent immersed in any one game is only a small fraction of the time they
spend engaged with the fan culture as a whole. This study will attempt to
honor that fact and contribute to the emerging scholarship that combines
gaming literature and fan literature, arguing that any discussion of the virtual
realm must also consider what happens *beyond* the virtual realm. We must
ask what is important to gamers as *fans*, participants in a cultural system
surrounding their favorite texts, genres, gaming-systems, and the act of gaming itself.

The specific aim of this chapter is to work through an especially pertinent
issue in the subcultural systems surrounding game enthusiasts: the troublesome issue of ownership. The issues arises because audience production both
in and *about* videogames is so prevalent and integral to the industry that the
line between producer and consumer is particularly susceptible to blur. First,
I will outline what fan studies can teach game studies about audience/produc-
er tension over ownership of a media text, using the research to expand on a
theoretical link I have made in the past: fan labor as New Organizational
labor.[7] Next, I will use this guidance as I explore one community of videogame fans with quite the story to tell when it comes to issues of ownership in
the production of the texts they love: *Fallout* fans. Implications will conclude
the study.

OWNERSHIP AND TEXTUAL POACHING IN FANDOM

It is important to foreground gamers as fans because this gives us an established history of scholarship to apply to the social and cultural behaviors that
occur both within and beyond the virtual realm. Likewise, it is important to
explore the ownership issue because of how ambiguous and fluid the "official product" is in gaming. The world of the gamer is a world where interaction with the avatar[8] is more customizable, responsive, and personal than
with the actor or musician, muddling at a fundamental level the split between
producer and consumer. This means that, for game fans, social interpretation
of a text takes on special significance. Because of this, Malaby proposed that
we must consider the social *process* behind games, one that occurs both
within and outside of the game world, one that "always contains the potential
for generating new practices and new meanings, possibly refiguring the game
itself."[9] Furthermore, thanks to technological advances, fan production in
game culture is also increasingly possible, and even increasingly noticed by
producers as fans labor for the texts they love.[10] All of this means that
ownership of fan labor is an especially intriguing issue to game studies.

Fan studies can teach game studies quite a bit about ownership. Going as far back as Jenkins' work in the early 1990s, fans have been thought of as "textual poachers," nomads traversing a textual landscape not their own, adapting it as best they can.[11] In such an environment, their unofficial investment does not lead to official ownership—control over the nature and destiny of the text they love. They must poach because producers maintain the rights and means to alter the text. Further, as fans poach, producers often attempt to exert control over their productions.[12] In this back and forth, fan investment in and production surrounding a media text can become so prevalent, organized, and entrenched that fans stop and ask "just whose text is this anyway?" Scardaville notes that often soap opera fans have been involved with a series longer than the revolving crew of actors, writers, and directors, prompting this same question when producers attempt to change its direction.[13] Producers, of course, own intellectual property (IP) rights to a media text and have traditionally controlled the means of production, but fans, the argument goes, keep those producers afloat and have immeasurable social and cultural investment in the text and so have some claim to ownership themselves.

The key idea behind textual poaching, I argue, is that fan ideas about "integrity" of the text (what is considered valid, consistent, accurate, or whole) must coexist with producer interests.[14] Tensions manifest when the two are perceived as at odds. To productive consumers, the intent of the producer is less important than the integrity of the narrative, so texts may rightly be appropriated by fans. As fans begin to understand a text in terms of their entitlement and ownership, they often must contend with their relative powerlessness in the face of producers who may alter it in ways the fans find unsuitable. No matter how invested fans are in a text, they have to acknowledge they are not its official owners and must approach the text, and its producers, as outsiders who can only indirectly influence its destiny. Because of this marginalized status, an essential part of being a fan is balancing loyalty and disappointment.[15] From there, it is up to the fan to theorize on ways to improve the text. This function is so prevalent that Jenkins calls fandom "first and foremost, an institution of theory and criticism."[16]

Such work toward text integrity can take diverse forms. Most basically, fans engage in *criticism* and *creativity* surrounding their favorite texts. Both are implicit attempts to poach a form of ownership from text producers, either by influencing producers or laboring without regard for them. Both criticism and creativity are important functions within fan culture. As Newman explains, "the task of the fan is a dual one that pushes and explores at the edge of the canon, expanding, modifying, enriching, while also preserving, policing, and remedying."[17] In all of this, fans do more than play. They engage in recreational labor. This can be formal production such as fan-fiction, mashups,[18] or fan reviews; it can also be labor that doesn't produce an artifact such as evangelizing the merits of the text to peers and petitioning networks to keep a program on the air.

Fan *criticism* occurs when fans publicly evaluate the integrity of a media text. Often, these criticisms occur when fans feel producers have mishandled a text, and wish to vocalize their objections to both other fans and producers. Some infractions include: a violation of the truth of the text's universe, a change in the text's aesthetic perceived as inaccurate or invalid, dislike of a certain character or plot, or even a complete dislike for an entire text. Criticism often occurs when fans feel producers have slighted them by picking profits over integrity or mass appeal over creativity (see Mihelich and Papineau's study of Jimmy Buffet fans and their struggle to deal with his increasingly contradictory bohemian message and corporate agenda).[19] Through such criticism, fans are arguing that they have just as much understanding of and insight into a text as producers, even if they must work as outsiders in order to accomplish change. When the criticism becomes intense enough, and when the perceived wrong becomes great enough, fans often mobilize, becoming organized and sometimes influential activists for a specific text. This is a practice as old as modern fandom, but the Internet has been especially helpful in giving these fan activism campaigns voice and reach.

Sometimes as part of this criticism, sometimes in response to failed criticism, and sometimes as a completely separate endeavor, fans often turn to their own *creativity* and productivity to work toward text integrity. This can occur when fan interests lie outside of the official text, or when fans feel the text merits further fleshing out that will not be done by producers. Whether it is fan fiction, fan art, or mashups, fan creativity supplements a text to make it more appealing or relatable to fans. In short, as Baym puts it, "fans transform their criticisms into opportunities to let their own creativity shine."[20] However, this production is not always in opposition to producer interests. It can also be very supportive of producer goals. Some groups of fans find a sense of ownership in coming alongside the official readings of the text and "filling them in" in sanctioned ways. For instance, Rehak discusses the productive and simultaneously supportive fans of the transmedia *Tomb Raider* character, Lara Croft, who channel their creativity toward producer-sanctioned goals.[21]

Digital-connectivity further complicates this issue, since now fans and producers are interacting on the Internet with increased regularity and consequence. Fans, while still often feeling marginalized, are at least feeling less invisible.[22] This increased communication and consequence is leading some to wonder whether the relationship between fans and producers is as one-way as it used to be. Even Jenkins has recently made a shift from his 1992 views that "like the poachers of old, fans operate from a position of cultural marginality and weakness" and "have only the most limited of resources" with which to influence producers.[23] Jenkins has since noted that active fansites (hubs of creative consumption where fans interact, create, and critique), and producer attentiveness to these sites, may be subtly altering the relationship between fan and producer.[24] The "fans as peasants" reality of fifteen years ago may be shifting, as discourse between fans and producers has experienced an internet-afforded boon.

Others, however, argue that this new era of interactivity may be the same old corporate wolf in an empowered sheep's clothing.[25] As Örnebring puts it:

> to be sure, media convergence is opening up new possibilities for interactivity, but it is difficult to ignore the fact that much of the interactivity on offer is produced by the "usual suspects" of transnational media conglomerates, and that audiences are addressed primarily as consumers or cultural artifacts.[26]

Tushnet argues that productive fans are well aware of this lack of authority and control. To Tushnet, "fans seem to see their legal status as similar to their social status: marginal and, at best, tolerated rather than accepted as a legitimate part of the universe of creators."[27] Any investigation over ownership in this era of digital-connectivity between producers and fans must be mindful of this tension.

POACHING, OWNERSHIP, AND LABOR IN GAME FANDOM

The negotiation of ownership between fan and producer is, I argue, especially salient in the ultra-customizable world of videogame fandom. Even though gamers poach through criticism and creativity just like other media fans, in gaming, the nature of the text is more interactive and emergent, and more ambiguous in regards to ownership issues than in any other media text.[28] Gaming, for purely technological reasons, also allows for more customization, which blurs the "whose text is this?" line. For instance, Lowood tells of *World of Warcraft*[29] players who bypassed a ban on chatting with enemies by using non-banned numerals and punctuation to get their message across, creating a specialized non-alphabetical language to use within the game.[30] Likewise, Grimes points out that ownership of the avatar itself has become a tricky legal issue in massively-multiplayer online games (called MMOs) since it is so customizable, personal to the player, and requires so much investment and energy to create and maintain.[31] Just who "owns" the avatar in an MMO (the producer who wrote the code or the player who put in the hours to implement the code) is an open question. In short, game texts can be extended and appropriated in more ways, with more success, than many other media texts.

"Mods"—an established form of this labor in game fandom—are a prime example here. In a mod (short for "modification") fans with programming prowess change the game at a code level, altering missions, stories, characters, and environment. Some mods build sequels or fix perceived incompletions within the title. Some are done alone, others are created socially in loose networks of laborers. Modding is popular enough that many games ship with a "construction set," a program that makes modding easier. Modding can be a very effective, successful, and aesthetically convincing form of appropriation.[32] If modding is the gamer equivalent of making a fan film that re-shoots a movie, this re-shoot could employ all the same actors, sets, and special effects as the original movie, since the modder has access to all the original code of the game. In all, gamer culture is a fan culture where "most radically put, the very product of the game is not constructed simply by the designers or publisher, nor contained within the boxed product, but produced only in conjunction with the players."[33]

Gamers are productive consumers outside of the virtual realm as well. "Machinima" films, those created by gamers using only in-game characters and graphics, are distributed and critiqued by fan communities outside of the game. Consalvo highlights the productivity of *Zelda*[34] fans, who work in collaborative networks to create "walkthroughs"—step-by-step guides to completing games and unlocking their secrets.[35] Demers outlines how *Dance, Dance, Revolution*[36] fans intersect game-fan production and music-fan production.[37] Pearce chronicles how the fans of a defunct MMO called *Uru* went into *Second Life*[38] and recreated the world of *Uru* using the space's customizability. The gamers built their own *Uru* island, complete with landmarks, puzzles, and characters from the game. In doing so, "players have quite literally taken it over and made it their own, carrying it forward to a new level."[39]

Gamer productivity is also an exemplary site of potential cooperation between fans and producers (or increased commodification, depending on the nature of the interaction). Like within broader media, gamers are seeing increased interaction with producers. The difference is that their interaction is finding a place of legitimate value within the videogame production system. Producers are increasingly reliant on fans to create user-generated content surrounding the game, broadening its networks of fans. Kücklich points out that mods, however unofficial are beneficial to producers in that they create an extension of the brand, add to the shelf life of the game, increase customer loyalty, and increase innovation within the industry (and even create a recruiting pool for the industry, allowing some to parlay their love into official careers within the industry).[40] Sotomaa discusses the official value of fan labor in relation to the movie-production simulator, *The Movies*;[41] Banks and Humphreys in relation to the railroad simulator, *Trainz.*[42] In an era where this labor is becoming more noticed (whether welcomed, rejected, or commodified), it is also becoming more consequential. Given that this is especially true of game fans, it is time to better understand how for game fans, the labor beyond the virtual realm compliments the play within it.

FAN LABOR AND THE NEW ORGANIZATION

All of this means one *must* think about fan labor and fan ownership in tandem. I have argued elsewhere[43] that we need a new paradigm to understand how productive consumers and the producers of media texts might interact in an age of digital connectivity.[44] The "New Organizational" paradigm[45] I adapted as a theoretical description of fan labor has been helpful in understanding fan productivity, but it can be expanded, in light of the unique nature of the ownership issue in fan studies, to include more on the subject. In the New Organization, production is in the hands of workers laboring across traditional organizational lines. With the affordances provided by information-communication technology, connectivity across broad geographical and hierarchical divides is fostered, innovation is demanded, and all organizations must be increasingly entrepreneurial to carve out a place in a narrowcast and cluttered market. The worker in the New Organization is primarily a "knowledge worker," dealing in information, connected in fragmented networks. The knowledge worker and the New Organization must primarily deal in "immaterial labor."[46] The New Organization is smaller and sleeker, more responsive and innovative. It must look beyond traditional organizational structures to succeed.

So what can a paradigm birthed out of organizational communication teach us about fans? I argue quite a bit. Take away the assumption that the member of the "New Organization" is a paid staffer, working full-time for their employer, and it becomes easy to see parallels between fan productivity and the employee in the New Organization. De Peuter and Dyer-Witheford have already made links between Hardt and Negri's immaterial labor and modders.[47] Terranova has argued that productive consumption is essentially a facet of a "gift economy" where compensation is "willingly exceeded in exchange for the pleasures of communication and exchange."[48] Producers, such as Electronic Arts (EA) with its franchise *The Sims*,[49] have long been relying on the innovation of fans working in extended networks to create content that drives the brand. And given that the primary social currency of fans is information and interpretation,[50] it is not too hard to envision them as knowledge workers, contributing in fragmented networks what they contribute best—immaterial labor. My own research[51] shows that *Fallout* fans viewed themselves as essential, if underappreciated, members of the organization: knowledge workers working for the text. But with this autonomous work comes tension over intellectual property.[52] If we want to use the New Organizational paradigm to foster increased cooperation between fans and producers, then the ownership issue must be better understood. The next section sets up an ideal place to study these issues: *Fallout* fandom.

THE LONG ROAD OF *FALLOUT* AND ITS FANS

Fallout 1 & 2 are role-playing games (RPGs)[53] that were released in 1997 and 1998 for the personal computer (PC) by the videogame publisher Interplay. They were developed for Interplay by Black Isle Studios. I became a fan of the games a few years later in 2001, after a friend and longtime fan introduced them to me. What met me when the CD first whirled to life in my computer's drive was an engaging and unique experience that I have revisited several times since. *Fallout 1 & 2* center on surviving and questing in a post-apocalyptic America circa year 2200, which has been devastated by a nuclear world war with ambiguous sides and no real victors. The story itself is just as gray. The player is not forced to be good, or even forced to choose between good and evil. The player is only asked to do what needs to be done to get by, creating a tense feeling of moral ambivalence that compliments the wasteland well. Federal and state governments have been replaced by city-state towns governed by criminal organizations, militias, martial law, or rudimentary legislative systems. In between these feuding city-states lay only anarchistic wilds. During each play through, it has been up to me and my player-character to tame these wilds or infuse them with chaos, save the world or sacrifice it, do our best to do good or do our best to survive.

I am not alone in my appreciation of *Fallout*. Soon after their release, the titles developed loyal groups of fans who deeply enjoyed the distinctive games.[54] Stylistically, mechanically, and creatively, *Fallout 1 & 2* have given fans like me much to celebrate. The player character is completely customizable, with gameplay consequences altered based on how the character is modified. The combat is turn-based,[55] which emphasizes strategic command and character competency over player reaction time or hand-eye coordination. The in-game perspective is isometric,[56] aiding in strategy. The game world is open-ended, with player choices deeply affecting character and game outcomes. Its dialogue is rich, with customizable character factors such as charisma, perception, and intelligence affecting a conversation's style and consequences. Finally, *Fallout* became known for its quirky brand of dark humor and a retro-future, 1950's, "wastepunk" *mis en scene* (*Mad Max* meets *The Jetsons*).

Soon, several *Fallout* fansites such as *No Mutants Allowed* (*NMA*) and *Duck and Cover* (*DaC*)[57] emerged and began engaging in all the typical fansite functions: posting screenshots, news, mods, walkthroughs, and fan fiction and art, as well as hosting forums for discussion. I never became a participant in any communal sense on any of these fansites, but I did engage their content often over the years, downloading patches and mods, lurking in forums, and checking for news updates. Through their productivity, the *Fallout* fans who did interact with each other became participants in, and advocates for, a fan culture that extended far beyond the games' borders. They believed in the game enough, and defended it with enough tenacity, to gain the reputation of being one of "gaming's fringe cults."[58]

As time went on, however, it appeared that *Fallout* fans would be without new installments of the game. Two more games based in the *Fallout* universe were released in 2001 and 2004 (*Fallout: Tactics* and *Fallout: Brotherhood of Steel* respectively), but neither were considered canon on fansites such as *NMA* and *DaC*. These two games broke drastically with many of the stylistic and gameplay qualities listed above that made *Fallout* so beloved by its fans, and also were narrative spin-offs, not continuing the main story of *Fallout 1 & 2*. The much-anticipated, and much-delayed, third installment of the primary *Fallout* narrative, *Fallout 3*, was cancelled in 2003 when serious financial trouble at Interplay forced staff reductions.[59] By 2005, plans for *Fallout 3* came to a screeching halt. *Van Buren*, the development name for *Fallout 3*, became a mythic item among fans, a Holy Grail, never realized.[60]

With the apparent death of their beloved series, despairing *Fallout* fans began to engage the text as if it was their own, picking up the torch and keeping the spirit of the game alive online. They had gone from marketing reps to consumer advocates in a game culture that had marginalized their interests. *NMA* administrator and fan opinion-leader Thomas "Brother None" Beekers had this to say:

> With the times, our goals have changed. Originally, we were formed to be supportive as we could be of *Fallout*, and this was great between *Fallout 1* and *2*, before *Tactics* release dashed our hopes of a good spin-off and no new release was forthcoming. . . . Now we're mostly evangelists of recreating the original *Fallout* experience. We try to convince the media and publishers that there is a viable niche market for *Fallout*-like games that has been underserviced for years.[61]

Eventually, the *Fallout* franchise was resurrected, but not by Interplay, and the next game would not be developed by Black Isle Studios. Instead, in April 2007, Bethesda Softworks acquired full legal ownership of the series and officially announced its control of the next sequel.[62] *Fallout 3* would be developed, but in an updated, mainstream style, with several beloved game elements changed. Moreover, Bethesda's Todd Howard, not Black Isle's

Brian Fargo, would be at the helm as executive producer of the game. Part of the game's re-imagining included updated graphics and sound, a switch to a first-person view,[63] real-time combat,[64] and a break from the central narrative of *Fallout 1 & 2*. *Fallout 3* had no major storyline connections to the original games. While these differences were enough to fundamentally alter the way the game played when I got my hands on it in October 2008, Bethesda had at least attempted to keep the mythos, wastepunk style, and even the dark humor of *Fallout 1 & 2* intact.

After Bethesda's announced takeover of the *Fallout* franchise, a debate surfaced within the active online *Fallout* fan community as to the quality of the next installment in the *Fallout* series. Skeptics of Bethesda's direction for the game have voiced several concerns. Bethesda's previous blockbuster fantasy game, *The Elder Scrolls IV: Oblivion*, is a starkly different RPG than *Fallout*, one fans contended is more action-oriented and childish.[65] The use of *Oblivion*'s graphics engine for *Fallout 3* gave rise to the "*Oblivion* with guns" protest across the fansites. Also, Bethesda was thought to be bad at dialogue and produced more-or-less linear plots without much choice or moral ambiguity. To many participants on the fansites, all of this meant a game made impure in an effort to reach a wider audience. They feared that as the gameplay, plot, combat system, and camera angle changed, so would the deeper things they loved about *Fallout*: its personality, humor, style, strategy, intricacy, and intelligence.

In contrast, supporters claimed that times are changing, games are evolving, and the original *Fallout* games were lacking in a few areas that Bethesda had the ability to update. It was not as if a 1998-style game would even be fathomable in a 2008 market. Even more, they believed that Bethesda had produced quality games in the past, and had always thoroughly considered and responded to player feedback when developing games. And furthermore, they argued, Bethesda could realistically do nothing to assuage a fan-base that had made up its mind over the game's poor potential as soon as it was announced. While the majority of vocal fans were skeptical about Bethesda's direction for *Fallout 3*, the debate occurred with great depth, breadth, and passion on many of the *Fallout* fansites, including Bethesda's official *Fallout 3* forum. Personally, I remain ambivalent about Bethesda's handling of *Fallout 3*. It had many strengths, but there were many times I stared at the screen thinking, "they did *this* to *Fallout*?" But I am not out to rule on which party is right and which is wrong. I am only here to investigate how fans and producers communicated diverse perspectives on ownership during an important historical moment.

Since *Fallout* fans responded to their text being abandoned in a way similar to the *Uru* fans highlighted earlier, poaching ownership of the game's destiny when no producer would, it becomes easier to understand why many of them were skeptical of Bethesda's re-re-appropriation of the text they appropriated as theirs. Their opinions are worth investigating as we consider ownership and labor issues in videogame fandom. I chose to perform this investigation on Bethesda's Official *Fallout* forums, asking: *how do fans active on Bethesda's* Fallout 3 *forum perceive their level of ownership of the game series?* After a discussion of methods, the question will be answered below.

METHODS

Research Site

The timing of this investigation is important because, as Jenkins points out, "the tension between the producer's conception and the fan's conceptions of the series are most visible at moments of friction or dispute."[66] Implicit ideas about ownership and labor become explicit during such tension. This tension was palpable on Bethesda's official forum space,[67] which contained forums devoted to the *Fallout* series. It seems that quite a few fans flocked to the Bethesda forum because it included a unique feature not present on the many fansite forums: developer presence and participation. Because of this potential for interactivity, Bethesda's official forum became a popular place to criticize and create, right under the noses of those who *did* own the game. It was the ideal place to conduct my research as I gathered data in the months prior to *Fallout 3*'s release. It was an active forum as well, containing a total of 53,485 posts on 1,264 subjects when data was collected, with "rolling" threads falling off after about six months.

During my time on the site, fans repeatedly discussed a pair of specific issues that were significant from an ownership perspective. First, a "Faction Profile" on the *Fallout 3* official site[68] about the Brotherhood of Steel, a joinable faction (a guild of sorts, where the player character could receive goods and get quests) from *Fallout 1 & 2*, raised considerable ire among fans. In the Brotherhood of Steel faction profile, *Fallout 3* Lead Designer Emil Pagliarulo detailed how the Brotherhood of Steel made it from California, the setting of *Fallout 1 & 2*, to Washington D.C., the setting of *Fallout 3*. Continuation of the narrative by Bethesda, which is neither a fan organization nor an original creator of the intellectual property, upset some fans who felt the company was treading where they should not, extending the narrative and robbing fans of the ownership they had claimed.

Second, the *Fallout* 10th anniversary contest[69] asked fans to design their own character "perk" (a choosable character customization that affects how the character functions in the game world). Winners received material prizes for their creativity. The grand-prize-winning perk was actually implemented into *Fallout 3*. This contest initially gave fans high hopes that their opinions were valid enough to merit consideration in the game development process. However, the chosen winner was met with frustration, and feelings that fans were being given a token PR event, not one where their opinions were really valued.

Analysis

My long history as a fan of the game series, as well as my familiarity with many of the fansites, gave me sufficient context to explore the discourse between fans on the official forum. When it came time to move beyond recreation into rigorous examination, I spent the better part of a year on both the official *Fallout 3* forum and other fansite forums, gaining a sense of what issues where important to fans and how issues important to fans emerged in their discourse. My participant-observer status allowed me to effectively understand the central issues and specialized language of the *Fallout* fan community. During data collection, I took a discourse analytic approach, pouring through the thousands upon thousands of posts on the official *Fallout 3* forum, looking for what the posters themselves had to say about their level of ownership in the game. This method was ideal because, as Lindlof, and Taylor point out, "public problems" like debates about policy (and, in this case, intellectual property and ownership policies) lend themselves to discourse analysis since "of particular interest to communication researchers is the role of discourse and other symbolic forms in the way in which conflictive issues are understood by participants and audiences."[70]

When it came to analysis, van Dijk's maxim that "discourse studies are about *talk and text in context*" guided my investigation.[71] Most fundamentally, I used pattern recognition in order to uncover structure and normative rules within mundane interaction.[72] Investigating these patterns, as well as deviant cases, helped provide the broadest sense of how *Fallout* fans and producers structured their interaction with each other. I paid attention to how fans spoke about their role within the organization, and what they felt they had to contribute. The threads collected and coded for this study are representative of a complex investigation, covering many issues, and those quoted were chosen because of their exemplary nature. A pilot study using similar methods was completed in July 2007, and data collection for the primary study began in September 2007 (a year before *Fallout 3* was set to release) and ran for six months until February 2008. As posts are quoted in the results and discussion section, their grammar and spelling are corrected only to improve readability.

RESULTS AND DISCUSSION

Poachers Still

Most fundamentally, and most important to foreground everything discussed below, fans on the official *Fallout 3* forum recognized that the text they esteemed was not theirs to control. Even in this era of digital connectivity, they were poachers, doing their best to criticize and create outside of official status within the game-development process. The game may have been theirs for a time, when Black Isle and Interplay had given up on it, and the hopes of *Van Buren* actually coming out were slim. But as Bethesda announced their plans for *Fallout 3*, it became evident to many fans that the text they loved so well had fallen out of their hands. They could only wait, pensive and skeptical about *Fallout*'s new direction. One fan used a hypothetical submission to the 10th anniversary perk contest to explain the plight of being a *Fallout* fan in *Fallout* terms (thread title and post number follow each post):

> *"Old Fallout Fan:" Living year after year in Fallout droughtiness has honed your patience but embittered your heart. You gain +25 to Survival, +1 to Endurance, but -1 to Charisma.* [73] ("10th anniversary contest 1," post 78)

So, beleaguered, powerless, and bitter, the fans continued to soldier on, resigned to the fact that they could not really do much to alter the game they loved so well. It was Bethesda who was changing things. The consensus of many fans was that Bethesda was "rebranding" *Fallout*, robbing it of its very heart right under the fans' noses:

> *Now, after reading this diary, I feel even more fear that my beloved franchise will end up pathos and epic. Recent design is just not like it should be IMO. Sorry, Bethsoft.* ("New diary today," post 172)

> *About the last thing hardcore fans have in mind is how to make money. Hence, about the last thing the producers and devs* [74] *have in mind is giving hardcore fans any kind of power to influence the development of a product.* ("*Fallout 3*: Need help?," post 131)

> *The point is moot. The train has already left the station. They never had any intention of maintaining fidelity to the basic principles of Fallout. Fallout, an RPG that went back-to-basics when everyone else was selling out, a game that's main influence was PnP RPGs.* [75] *They like the setting and want to jam it into the Oblivion cookie cutter mold to make more money.* ("*Fallout 3*: Need help?," post 18)

A fan-generated poll in the "*Fallout 3*: Need help?" thread asked if hardcore fans could influence the game development process. In the straw poll, only 18 of 123 respondents said "yes." Eleven said, "I don't know," Forty-six said "no," and 48 said, "I hope so." The vast majority of posters in the thread were clear on one point: that fans were not the ones who decided whether they had any ownership of the series. It was the producers who could grant or withhold such a privilege. Fan perceptions of influence over text integrity were clear on this point:

> Beth,[76] *for better or worse, has its own ideas and direction on how this game will develop. Let's be realistic guys. Posts on a forum are not going to influence this direction. The best any of us can hope to do is come up with some relatively minor detail Beth overlooked that they deem worthy to throw in.* ("*Fallout 3*: Need help?," post 88)

Now, not every fan active on the *Fallout 3* forum felt that Bethesda was doing a poor job as they produced the game. But even in these cases, a consistent pattern emerged. Whether coming from a pro- or anti-producer perspective, most fans used "they" and "you" language when referring to *Fallout 3* and Bethesda, indicating that what "they" were doing and what "they" planned for the game was out of fan control. Bethesda was the "Other" who owned the game. Fans were the distanced and marginalized, who did not have the influence to alter the course of the game, save for the tactics of the poacher or protester. Several examples illustrate this perception:

> *They decided to do Oblivion with guns, instead of doing Fallout 3. Terrible decision in my opinion. In every aspect.* ("*Fallout 3*: Need help?," post 8)

> *This is an RPG with a history and lore of several centuries, so it's important that their writers take some time to clear up all the missing info and previous contradictions in Fallout lore so that we can enjoy Fallout 3.* ("What the Brotherhood of Steel really is," post 139)

> *You tell me it's Fallout it better damn be Fallout. Man false advertisement is illegal.* ("What I fear the most about *Fallout 3*," post 21)

Bethesda was the one that could change the text. Bethesda was the one that controlled official canon. The very fact that they had chosen the name *Fallout 3* for the game communicated as much. Several fans indicated that they would have been less displeased with Bethesda making yet another non-canon *Fallout* spin-off, but a name like *Fallout 3* came with an air of canon, and therefore would be harder to disavow, even if inferior. *Fallout* fans, utopian claims aside, saw themselves as outsiders within the system of production surrounding *Fallout 3*.

The *Fallout 3* situation adds an interesting twist to this ownership struggle, however. In the case of *Fallout 3*, even if fans believed that the game was not theirs to control or alter, they were not quite sure that is was Bethesda's either. There were more than a few accusations that Bethesda hijacked intellectual property from Interplay and Black Isle, the text's true owners:

> *Bethesda bought a piece of paper that allows them to do whatever they want and say it's justified.* ("What the Brotherhood of Steel really is," post 16)

> *"But doesn't that change the continuity of the originals?" "Hello? We have the IP now. We can do anything we want!"* ("My view on *Fallout 3*," post 8)

> *The name says Fallout 3, but the PR says, to an astonishing degree, that "we're doing things our own way. This baby's ours now, the past be damned."* ("What will it take for you to buy *Fallout 3*?," post 29)

To many fans, this was the heart of the complaint about rebranding. The text was not Bethesda's to rebrand, and many felt that Bethesda was not rebranding it well. Bethesda's chosen perk winner in the 10th anniversary contest was mostly viewed as unsatisfactory, since it lacked either trademark *Fallout* humor or a way to be implemented without negatively altering gameplay. Their vision of the Brotherhood of Steel was also considered invalid, thought to be either high-fantasy influenced like *Oblivion*, or following the simplistic narratives of *Tactics* and *Brotherhood of Steel*. This is an especially damning charge since *Fallout* fans treated an invalidation of the established truth of the text with the utmost of contempt. As Baym said of soap fans dealing with the same re-negotiation of canon, "one flaw that is rarely funny is violation of the truth of the fiction established through prior shows."[77] One fan powerfully articulated these objections:

> *They're not fans of what Fallout stood for. They're making a cult game into a vapid mainstream fare. Fallout was the total antithesis to what Bethesda does now. Making Fallout into an action game with stats would make it a total anathema to what it stood for. Big fans my tookus.* ("*Fallout 3*: Need help?," post 18)

One fan displayed an altered forum signature, parodying the slogan of the first two games, "*Fallout*: A Post-Nuclear Role Playing Game," to comment on all the alterations that were occurring. The poster altered the slogan to read "Fallout 3: A Post Nukular **Action** Role Playing Game." The satire is telling. The word "post" is removed and only "nuclear" remains, a commentary on the nuclear explosions rumored to be rampant in *Fallout 3* (which were more toned-down in the first two). "Nuclear" is misspelled in a manner fitting a dumbed-down game. "Action" is added in bold, reflecting the feared tone of *Fallout 3*. The consensus among the skeptical seemed clear. Despite the best of fan efforts, *Fallout 3* was being hijacked by a company that knew and cared little about its heart. This position is articulated expertly below:

> Even before Bethesda was in the picture, NMA, DaC, and the Codex[78] have a long consistent history of clearly stating what they expect a sequel to be. Years. And years. Consistently. The blueprint was there as laid out by the original devs, and the flame kept lit by the faithful. It's Bethesda who has usurped ownership and is imposing its changes based on its (commercially) successful yet shallow formula. Duck and Cover, NMA, Codex, they've never changed their tune. Their vitriol is only in response to Bethesda's negligence. ("*Fallout 3*: Need help?," post 100)

(Implicit) Ownership and Justified Activism

But this feeling of powerlessness is not the whole story of the ownership issue on the *Fallout 3* forum. Because of the perception of being hijacked, perhaps, indications of implicit ownership were prevalent among fans. It was a type of ownership that seemed to say, "even though I can't control the destiny of this text, I certainly have the right to tell you all about how it should go." Activism was seen as justified in light of the circumstances, and this is the "*why*" to answer the "*how*" of fans' New Organizational labor. Fans labored in fragmented networks of knowledge workers in order to exercise an implicit ownership over *Fallout 3*. This lines up with my previous findings that fans were more interested in altering the *official* text than altering it *unofficially* through mods.[79] Feeling that they had a right to some ownership in the process, fans chose to labor more officially than unofficially, at least as long as *Fallout 3* was still being developed.

Many fans were near-parental in their suggestions for Bethesda, offering tidbits on how they would create the text if indeed they could. Immense detail and full threads were devoted to everything from the in-game radio to weapons to combat to character creation. Perks suggestions, both before and after the contest (including the ubiquitous claim that "mine should have won"), indicated a feeling of implicit ownership. The Brotherhood of Steel faction profile posted to the official site prompted a myriad of "it should have gone like this" or "why didn't this happen?" suggestions to correct Bethesda's choices. The response was so strong, in fact, that the fansite *NMA* ran a contest inviting fans to write their own accounts of how the "Brotherhood of Steel" made it to the east coast and what they were doing there. Fans quoted descriptions from a popular *Fallout* fan wiki site, called *The Vault*[80] to argue what the Brotherhood really should be, countering what Bethesda proposed about them. To one poster, what Bethesda thought of the Brotherhood of Steel did not really matter once the game disc began to spin (at that point, Bethesda's control would be less complete):

At the end of the day, my frustrations aside, I guess I don't care. I will kill the Brotherhood soldiers at every opportunity, as the splinter group is a do-gooder "knightly" police force, and the original Brotherhood was a group of xenophobic racist zealot dictators. I've got a hot .45 slug ready for each and every one of 'em. DEATH TO THE BROTHERHOOD OF STEEL. ("What the Brotherhood of Steel really is," post 181)

The wide array of screen names, avatars, art, and so on dealing with the *Fallout* universe to be found on the forum all implicitly communicated fan experience and expertise to producers as they advocated their implicit ownership of the text. All in all, the situation was something of a paradox: powerless fans attempting to exert control over a text that they did not own, but they knew in their heart was truly theirs. This was the ownership struggle of the *Fallout* fan.

Despite this norm of implicit ownership, there did exist a bit of schism as to whether fans *should* be in control of the text, even if they could. There were those who said that fans did not have the right or ability to make any decisions for the game (as the poster directly below did when speaking about *The Vault* wiki site):

What I'm saying is that you can't really consider it canon. Not when anyone can put anything in it (something that would make canon a matter of majority rule, not a matter of the intentions of the games' creators). ("What the Brotherhood of Steel really is," post 48)

I greatly object to "hardcore fans"[81] *being in charge of a game's developments, because hardcore fans are opinionated pricks, plus a lot of people just don't have a clue . . . hardcore fans shouldn't assume they know better than everyone else.* ("*Fallout 3*: Need help?," post 48)

Still others did not mind that the text was out of their hands:

I've accepted the fact that Bethesda has the rights to make this game. I've accepted the fact that it's probably our only chance to get a Fallout 3. I've accepted the fact that no matter how much I sit at home in front of my computer and complain about what they're doing I can't do anything about it. And I've accepted the fact that it might be a great game anyway, so there's no reason to complain. ("My view on *Fallout 3*," post 20)

As these cases demonstrate, no matter what they thought about this implicit ownership, fans recognized that since they were outsiders any control over text integrity could only come through indirect suggestion, poaching, or from what was ceded by producers. Fans could challenge conventions in threads centering on perk ideas all they wanted, fudging on official contest rules for the sake of showing off expertise, making inside jokes, or playing with the text. However when these fans actually were serious about submitting a perk to be used within the game, they were deeply interested in Bethesda's wishes. Clarifications from producers in the threads referred to official contest rules, indicating that fans were only allowed to officially operate within the parameters set up by Bethesda. Indicative of this marginalization and inability to directly influence producers was the "I hope" language many fans used when contemplating various elements of the game. These posts ultimately demonstrate just how *implicit* fans felt their ownership was. There in spirit; not in consequence:

> *As for writing . . . one can only cross one's fingers.* ("What will it take for you to buy *Fallout 3*?," post 22)

> *I pray FO3 will not feature a jumping button.* ("10th anniversary contest 1," post 112)

> *It also seems somewhat lame to me. I was hoping they would pick something funny and creative that fit more with Fallout's original perks.* ("Your thoughts on the winning perk," post 16)

Criticism and Creativity: Pressing Hard and Walking Away

When seeking to understand exactly *how* fans active on the official *Fallout 3* forum acted on their feelings of implicit ownership, the criticism and creativity tactics addressed in fan literature provide an excellent classification system. Through these tactics, fans—working from the de facto position of outsider—made their boldest and most direct claims to implicit ownership of the text. Some did this by seeking to alter the official text, pressing hard against Bethesda. Others, however, walked away, and preferred to channel their implicit ownership toward unofficial labor.

Whether aimed at the official or unofficial, creativity was everywhere on the *Fallout 3* forum, as fans responded to and expanded on the information they were receiving from Bethesda. The various weapons ideas, perk ideas, and adjustments to the Brotherhood of Steel narrative became the realm of fans, who produced both out of pure recreation and perceived obligation. Some of this creativity was from a place of surrender, of channeling implicit ownership to more speculative and recreational ends. For instance, the 10th anniversary contest threads went well past the contest deadline and the announced winners. Fans continued to submit perks, many of them twisting the rules of what a perk officially could be, as far as the contest was concerned. They included anti-perks (negative effects a character might obtain), traits (starting characteristics that often have a positive and negative character consequence), and complex histories and descriptions. When one fan called others on their impropriety in the eyes of the contest rules, the response was quick:

> *It's probably a bad thing as they are not straight perks. Still, makes 'em more interesting to read here.* ("10th anniversary contest 1," post 162)

The underlying message seemed to be that "the game isn't ours but this thread is." One member listed 22 perk ideas in a thread that began after the contest was over. The same occurred in response to the Brotherhood of Steel faction profile as fans wrote up their own accounts of how the faction got to the east coast in their version of the *Fallout* universe (and many of those probably submitted their ideas to *NMA*'s contest asking for such accounts). One member even re-wrote and posted Lead Designer Emil Pagliarulo's intro to the Brotherhood of Steel faction profile and invited fans to critique it. And like other modders studied before them, those who advocated modding *Fallout* did so to produce a variant of the game they could enjoy, even if they could not enjoy the official text. Through all this, fans produced for their own implicit ownership, not for official inclusion. As fans posted their imaginative ideas for what the game could be, they were clear that they were not concerned about the annoying specifics of how these ideas would be implemented. In short, they poached. They knew the official text was not theirs, but that did not mean their own unique variations of the text's universe could not be theirs. In line with these feelings, many fans seemed to understand their creative manipulations of the text as unofficial and therefore inferior, agreeing with Tushnet's counterintuitive findings about fan empowerment in an era of fan-producer connectivity. [82]

While these creative diversions were common, it seemed that a majority of fans were also interested in influencing the official text. Many chose to press hard against Bethesda to alter *Fallout 3*. Since they were marginalized outsiders, criticism was the only method available to achieve that goal. If fans were unsure about their ownership status, they were certain criticism was one of their primary roles. And in this function was their most direct power. As many fans pointed out, they were the ones who got to call out producer inconsistencies, and ultimately decide whether or not to buy game, even if they did not get to help create it:

> *Don't buy games that suck. Otherwise, expect more of the same since by subsidizing them, you're encouraging more of the same.* ("What will it take for you to buy *Fallout 3*?," post 90)

*If I am going to spend good money on a game with "FALLOUT" in the title, it
better be good! Because if it sucks, I will be all over this forum with comments
about how I got screwed over. . . . YOU HAVE BEEN WARNED!!!!!!* ("What
will it take for you to buy *Fallout 3*?," post 71)

Fans often raised their voices against the transgressions they were otherwise
powerless to alter. For instance, this fan proposed a perk that would fix the
errors Bethesda was making as they developed the game:

*"Retro Boy Perk:" Level 1. Choosing this perk allows you to play whole game
from beginning to end in turn based mode with isometric view and story
comparable to the ones in original Fallouts. Perk is available only on PC
version of the game.* ("10th anniversary contest 1," post 134)

Many used criticism to express their dismay at how little they could actually
do to save *Fallout*, harkening back to their powerlessness, as this fan did in a
post addressed directly to producers:

*And who cares about the stupid anniversary. Fallout is dead. I suggest you
place this games' rights in a safe and never open it again. Go dig up a hole
somewhere and toss the darn thing in it so we can all get on with our lives.*
("10th anniversary contest 1," post 25)

However, not all criticism involved disparaging critiques. Some was level-
headed:

*I really liked the faction profile. One minor thing that troubles me—how would
the East Coast Brotherhood have communicated with the West Coast Brotherhood?* ("New diary today," post 15)

Even so, the goals of such criticism remained consistent. The integrity of the
official text and implicit ownership of that text were paramount ends.

Its recreational or unofficial status aside, forum-posting was generally
serious business to many *Fallout* fans. Or at least it had important goals, even
though many fans felt that they were too small to successfully affect any
deep change. After all, their wants, as they perceived the situation, were in
direct opposition to what would make the game profitable and appealing to a
mass audience. However, they soldiered on for the sake of the text they
loved. Some did so with near-parental concern, others with beleaguered
hopelessness, fewer with confident anticipation that Bethesda would honor
their wishes and create a text with integrity. All tactics acknowledged indirect influence at best and marginalization at worst. This is the final answer to
the ownership question, and a surprising one given some utopian hopes in
this era of digital connectivity. *Fallout* fans recognized that *Fallout* was not
theirs and that any influence over its destiny had to be poached or ceded by

producers. Fans used creativity to alter the text unofficially and criticism to attempt to alter it officially. The two functions were not perceived equally though. In all, the most attention and the most esteem, was placed on criticism, as the integrity of the official text became the central issue in many debates. Despite diverse opinions about whether *Fallout 3* could be saved by fans, most fans felt obligated to try. This was how they would own a piece of the game:

> *But because of people like us on the forums stating our opinions over and over, it pushes the company to actually develop the game the way the gamers want it to be and not the devs way. Of course they could just go the devs way but at least we would have tried to save the franchise.* ("What I fear the most about *Fallout 3*," post 32)

> *I don't know if there is a preceding moment in the history of game development when base fans could have changed the course of how a particular game was made.* ("*Fallout 3*: Need help?," post 1)

> *Fallout 3 MUST be like Fallout . . . the best answer for every question on this forum besides "I have the holy sacred duty to watch over my beloved game."* ("What will it take for you to buy *Fallout 3*?," post 1)

IMPLICATIONS

This study should be just one more piece of evidence that gamers *are* fans (or, better put, there are fans among game audiences just as there are fans among music, film, and television audiences). Game fans are productive consumers who take part in an entire "gift economy" that functions in interdependence with the official releases game studios offer. They criticize and create, just like other fans, and what's more their criticism and creativity has quite a lot of consequence thanks to the emergent, customizable, and interactive nature of the videogame industry. If this criticism and creativity is labor organized in a New Organizational network of knowledge workers, then we can now better say that the goal of this labor as exerting implicit ownership over the text, even if it is not granted officially by media producers.

This leads to realization that, as seems to happen so often with utopian predictions, the democratization of producer-consumer interaction in an era of digital-connectivity may have been overstated. While productive consumers have made great strides since Jenkins declared them "peasants" in 1992,[83] it is perhaps a bit early to assume that producers are declaring that "fandom is beautiful," even if scholars such as Gray, Sandvoss, and Harrington are.[84] More likely, producers are recognizing that "fandom is unavoidable" when looking at the habits of their target consumers, and that "fandom is profitable" when properly commodified. The official *Fallout 3* forum demonstrated as much. It was a place where powerless fans attempted to exert control over a text they did not own but they knew in their hearts was truly theirs.

As fans like me continue to engage in the social production of labor and leisure with increasing consequence, it is of increasing importance that we consider what happens *beyond* the virtual realm. As scholars theorize on the utopic dreams and apocalyptic fantasies that are part and parcel of being a gamer, we must consider how social and practical connection with other gamers and producers of games play into that experience. This study has found that, while gamers have increased opportunity to engage with the production process surrounding their favorite texts, the situation is not as simple as mere utopia. In all, to gamers, the troublesome issue of ownership only looks as if it will get more complicated—and consequential—as the years march on.

NOTES

1. *Madden* refers to the venerable National Football League game series put out yearly by Electronic Arts, which simulates professional American football.
2. *Rock Band* is a music simulation game where players don plastic guitars, basses, and drums (along with a less-plastic microphone) and play along with pop and rock music, karaoke style.
3. "Partying up" in the first-person shooter series *Gears of War* consists of creating a team of up to five players to compete online against other teams of five. Matches are decided by the last team standing.
4. See the following websites: http://www.ign.com, http://www.gamespot.com, and http://www.gametrailers.com.
5. See Jeroen Jansz and Lonneke Martens, "Gaming at a LAN Event: The Social Context of Playing Video Games," *New Media & Society 7*, 3(2005): 333–355; and Torill Elvira Mortensen, "Wow Is the New Mud: Social Gaming from Text to Video," *Games and Culture 1*, 4(2006): 397–413.
6. Garry Crawford and Jason Rutter, "Playing the Game: Performance in Digital Game Audiences." In *Fandom: Identities and Communities in a Mediated World*, ed. by Jonathan Gray, Cornel Sandvoss, and C. Lee Harrington (New York: New York University Press, 2007), 271–281; Henry Jenkins, *Convergence Culture: Where Old and New Media Collide* (New York: New York University Press, 2006); Henry Jenkins, *Fans, Bloggers, and Gamers: Exploring Participatory Culture* (New York: New York University Press, 2006); James Newman, "Playing (with) Videogames," *Convergence: The International Journal of Research into New Media Technologies 11*, 1(2005): 48–67; James Newman, *Playing with Videogames* (London: Routledge, 2008); and T.L. Taylor, *Play between Worlds: Exploring Online Game Culture* (Cambridge, MA: MIT Press, 2006).
7. R.M. Milner, "Working for the Text: Fan Labor and the New Organization," *International Journal of Cultural Studies 12*, 5(2009): 491–508.
8. The "avatar"—or often "player character"—is simply the object the player controls as the gameworld is traversed. This can be in the singular (like a person or a monster or even a paint brush), or it can exist in the plural (like one player controlling a battalion of tanks).
9. Thomas M. Malaby, "Beyond Play: A New Approach to Games," *Games and Culture 2*, 2(2007): 102.

10. See John Banks and Sal Humphreys, "The Labour of User Co-Creation: Emerging Social Network Markets?" *Convergence: The International Journal of Research into New Media Technologies 14*, 4(2008): 401–418.

11. Henry Jenkins, *Textual Poachers: Television Fans and Participatory Culture* (New York: Routledge, 1992).

12. Simone Murray, "'Celebrating the Story the Way It Is': Cultural Studies, Corporate Media and the Contested Utility of Fandom," *Continuum: Journal of Media & Cultural Studies 18*, 1(2004): 7–25.

13. Melissa C. Scardaville, "Accidental Activists: Fan Activism in the Soap Opera Community," *American Behavioral Scientist 48*, 7(2005): 881–901.

14. See R.M. Milner, "Negotiating Text Integrity: An Analysis of Fan-Producer Interaction in an Era of Digital-Connectivity," *Information, Communication & Society 13*, 5(2000): 722–746.

15. Nancy K. Baym, *Tune in, Log On: Soaps, Fandom, and Online Community* (Thousand Oaks, CA: Sage, 2000).

16. Jenkins, *Textual Poachers*, 86.

17. Newman, "Playing (with) Videogames," 53.

18. "Mashups" involve the remixing of multiple texts into a single text. For instance, cutting music and clips from visual media into a sort of music video is a common fan practice (as one example, there are quite a few mashups featuring the love story of Jim and Pam from the TV series *The Office* set to music on *YouTube*).

19. John Mihelich and John Papineau, "Parrotheads in Margaritaville: Fan Practice, Oppositional Culture, and Embedded Cultural Resistance in Buffett Fandom," *Journal of Popular Music Studies 17*, 2(2005): 175–202.

20. Baym, *Tune in, Log On,* 105.

21. Bob Rehak, "Mapping the Bit Girl: Lara Croft and New Media Fandom," *Information, Communication, & Society 6*, 4(2003): 477–496.

22. See Mark Andrejevic, "Watching Television without Pity: The Productivity of Online Fans," *Television & New Media 9*, 1(2008): 24–46; and Jonathan Gray, Cornel Sandvoss, and C. Lee Harrington, "Why Study Fans?" In *Fandom: Identities and Communities in a Mediated World*, ed. by Jonathan Gray, Cornel Sandvoss, and C. Lee Harrington (New York: New York University Press, 2007), 1–16.

23. Jenkins, *Textual Poachers,* 26.

24. Henry Jenkins, "The Future of Fandom." In *Fandom: Identities and Communities in a Mediated World*, ed. by Jonathan Gray, Cornel Sandvoss, and C. Lee Harrington (New York: New York University Press, 2007), 357–364.

25. See Andrejevic, "Watching Television without Pity"; Victor Costello and Barbara Moore, "An Examination of Audience Activity and Online Television Fandom," *Television & New Media 8*, 2(2007): 124–143; and Scardaville, "Accidental Activists."

26. Henrik Örnebring, "Alternate Reality Gaming and Convergence Culture: The Case of *Alias*," *International Journal of Cultural Studies 10*, 4(2007): 450.

27. Rebecca Tushnet, "Copyright Law, Fan Practices, and the Rights of the Author." In *Fandom: Identities and Communities in a Mediated World*, ed. by Jonathan Gray, Cornel Sandvoss, and C. Lee Harrington (New York: New York University Press, 2007), 60.

28. Sal Humphreys, "Productive Players: Online Computer Games' Challenge to Conventional Media Forms," *Communication and Critical/Cultural Studies 2*, 1(2005): 37–51.

29. *World of Warcraft* is an oft-studied massively-multiplayer online (MMO) game that is somewhat notorious for its intense time commitment and seemingly never ending supply of quests. Despite often being labeled as losers or loners, *Warcraft* players cannot function at the game's higher levels without social coordination and group work.

30. Henry Lowood, "Storyline, Dance/Music, or PvP? Game Movies and Community Players in *World of Warcraft*," *Games and Culture 1*, 4(2006): 362–382.

31. Sara M. Grimes, "Online Multiplayer Games: A Virtual Space for Intellectual Property Debates?" *New Media & Society 8*, 6(2006): 969–990.

32. Hector Postigo, "Video Game Appropriation through Modifications: Attitudes Concerning Intellectual Property among Modders and Fans," *Convergence: The International Journal of Research into New Media Technologies 14*, 1(2008): 59–74.

33. Taylor, *Play between Worlds*, 126.

34. *Zelda* is a classic adventure-game franchise that resides on many Nintendo systems.

35. Mia Consalvo, "*Zelda 64* and Video Game Fans: A Walkthrough of Games, Intertextuality, and Narrative," *Television & New Media 4*, 3(2003): 321–334.

36. *Dance, Dance, Revolution* is a music and rhythm game that consists of stepping on colored pads in time with music.

37. Joanna Demers, "Dancing Machines: 'Dance, Dance, Revolution,' Cybernetic Dance, and Musical Taste," *Popular Music 25*, 3(2006): 401–414.

38. *Second-Life* is the much-studied online social space with a 3D, game-like interface where users can create, buy, and sell virtual property, as well as socialize and engage with others.

39. Celia Pearce, "Productive Play: Game Culture from the Bottom Up," *Games and Culture 1*, 1 (2006): 23.

40. J. Kücklich, "Precarious Playbour: Modders and the Digital Games Industry," *Fibreculture*, 5 (2005), accessed at http://journal.fibreculture.org/issue5/kucklich.html.

41. Olli Sotamaa, "Let Me Take You to *the Movies*: Productive Players, Commodification, and Transformative Play," *Convergence: The International Journal of Research into New Media Technologies 13*, 4(2007): 383–401.

42. Banks and Humphreys, "The Labour of User Co-Creation."

43. Milner, "Working for the Text."

44. I am, of course, not the only one making this claim. See, for instance, T.L. Taylor, "Beyond Management: Considering Participatory Design and Governance in Player Culture," *First Monday* (2006), accessed at http://www.uic.edu/htbin/cgiwrap/bin/ojs/index.php/fm/article/view/1611/1526.

45. See Peter F. Drucker, "The Age of Social Transformation," *Atlantic Monthly*, November, 1994: 53–71; Peter F. Drucker, "The Coming of the New Organization." In *Harvard Business Review on Knowledge Management*, ed. by Havard Business School Press (Cambridge, MA: Harvard University Press, 1998), 1–19; and Janet Fulk and Gerardine DeSantis. "Electronic Communication and Changing Organizational Forms," *Organization Science 6*, 4(1995): 337–349.

46. To use a term advanced in Michael Hardt and Antonio Negri, *Multitude: War and Democracy in the Age of Empire* (New York: Penguin, 2004).

47. Greig de Peuter and Nick Dyer-Witheford, "A Playful Multitude? Mobilising and Counter-Mobilising Immaterial Game Labour," *Fibreculture*, 2005, accessed at http://journal.fibreculture.org/issue5/depeuter_dyerwitheford.html.

48. Tiziana Terranova, "Free Labor: Producing Culture for the Digital Economy," *Social Text 18*, 2(2000): 48.

49. *The Sims* is a simulator game franchise that puts the player in command of a household of avatars. The player must see to it that the avatars get what they need to survive (food, shelter, income, etc.), and, beyond that, also conducts their social life with intricate detail. *The Sims* is well-known for its fan productivity and producer support of that productivity. Characters, items, and architecture are all created by *Sims* fans and shared online.

50. Baym, *Tune in, Log On*.

51. Milner, "Working for the Text"; Milner, "Negotiating Text Integrity."

52. See Murray, "'Celebrating the Story the Way It Is'"; and Mia Consalvo, "Cyber-Slaying Media Fans: Code, Digital Poaching, and Corporate Control of the Internet," *Journal of Communication Inquiry 27*, 1(2003a): 67–86.

53. A *Role-Playing Game* (RPG) is a genre of videogame that is so called because it invites the player to take on and manage a role *within* the gameworld, rather than emphasizing player skill outside of the game world. While the term is ambiguous (and always shifting), RPGs are traditionally known for their increased level of intricacy, their customization ability, and their emphasis on tactical prowess over hand-eye coordination.

54. Joe Blancato, "Gaming's Fringe Cult" (*The Escapist*, June 19, 2007), accessed at http://www.escapistmagazine.com/articles/print/565); and Michael Zenke, "From Black Isle to Bethesda" (*The Escapist*, May 22, 2007), accessed at http://www.escapistmagazine.com/articles/print/545.

55. In "turn-based" combat (as opposed to "real-time" combat), the game plays more like chess and less like, say, paintball. Each "side" in a combat scenario is given a turn to inflict damage, or change positions. During this time, the other "side" cannot move or inflict damage. The result is a slower (and some say more strategic) combat process.

56. An "isometric" view of the field of play (as opposed to "first person") is a "bird's eye" view that allows the player to see it from above. Avatars are smaller, and the player gets a look at a larger portion of the environment when moving and fighting.

57. See the following websites, http://www.nma-fallout.com and http://www.duckandcover.cx.

58. Blancato, "Gaming's Fringe Cults."

59. Andy Chalk, "Interplay Crosses Fingers for Fallout MMOG" (*The Escapist*, August 15, 2007), accessed at http://www.escapistmagazine.com/news/print/75978); and Zenke, "From Black Isle to Bethesda."

60. Andy Chalk, "*Van Buren* Demo Released" (*The Escapist*, May 4, 2007), accessed at http://www.escapistmagazine.com/news/print/71437.

61. Blancato, "Gaming's Fringe Cults."

62. Andy Chalk, "Bethesda Softworks Purchases Fallout IP" (*The Escapist*, April 13, 2007), accessed at http://www.escapistmagazine.com/news/print/70622.

63. In a "first person" view the player "sees" only out of the avatar's eyes, as opposed to an isometric view where the player sees the whole field of play. This is said to make the experience more visceral, but less strategic.

64. "Real-time combat" (as opposed to turn-based) is combat in which both "sides" of the struggle can move and inflict damage at the same time. This is said to emphasize player hand-eye coordination and reaction time over strategy and character skill. To be fair, *Fallout 3*'s combat system is actually "real-time with pause," meaning that combat takes place in real time, but the player can pause the battle to execute focused shots and administer health and power-ups.

65. *The Elder Scrolls IV: Oblivion* shares with the first two *Fallouts* a rich world and a customizable character that strongly influences how the game is played. It differs in its more linear plot, scanter dialogue, and lack of game-changing choices.

66. Jenkins, *Textual Poachers*, 132.

67. See the following website, http:// www.bethsoft.com/bgsforums.

68. Emil Pagliarulo, "Faction Profile—the Brotherhood of Steel" (Fallout, n.d.), accessed at http://fallout.bethsoft.com/eng/vault/diaries_diary3–1.09.08.html.

69. Anonymous, "Fallout 10th Anniversary Contest" (Fallout, n.d.), accessed at http://fallout.bethsoft.com/vault/falloutcontest.html.

70. Thomas R. Lindlof, and Bryan C. Taylor, *Qualitative Communication Research Methods, Second ed.* (Thousand Oaks, CA: Sage, 2002), 73.

71. Teun A. van Dijk, "The Study of Discourse." In *Discourse as Structures and Processes*, ed. by Teun A. van Dijk (London: Sage, 1997), 2.

72. See Karen Tracy, "Discourse Analysis in Communication." In *The Handbook of Discourse Analysis*, ed. by D. Schiffrin, D. Tannen, and H.E. Hamilton (Malden, MA: Blackwell Publishers, Ltd., 2001), 725–749; and Teun A. van Dijk, "The Study of Discourse." In *Discourse as Structures and Processes*, ed. by Teun A. van Dijk (London: Sage, 1997), 1–34.

73. "Survival" here might be referring to "outdoorsman," a skill the player-character can master in the first *Fallout* games that relates to the ability to navigate its harsh terrain. "Survival" as far as I know, is never an official skill in the *Fallout* series. "Endurance" and "Charisma" are traits the player can invest in at the beginning of the game which relate to enduring damage and charming characters in the game respectively.

74. "Devs" is often shorthand for "developers"—a catchall for the staff working on the production of a videogame.

75. "PnP" most often means "pen and paper"—referring to the lineage of videogame RPGs. Pen and paper RPGs (such as the famous *Dungeons and Dragons*) require players to invent characters and roll dice as they move through a world created solely of their own imaginary pariticpation.

76. This was common shorthand for "Bethesda" on the forum.

77. Baym, *Tune in, Log On*, 99.

78. "The Codex" might be referring to an RPG fansite called *RPG Codex*, accessed at http://www.rpgcodex.net/. While not *Fallout* specific, the site does act as a sort of watch dog and advocate for the integrity of RPGs.

79. Milner, "Working for the Text."

80. See the following website, http://fallout.wikia.com.

81. "Hardcore" is often a label applied (and self-applied) to gamers who take their hobby seriously.

82. Tushnet, "Copyright Law, Fan Practices, and the Rights of the Author."

83. Jenkins, *Textual Poachers.*

84. Gray, Sandvoss, and Harrington, "Why Study Fans?"

Conclusion

Apocalyptic Fantasies and Utopic Dreams Untold —
Where Do We Go from Here?

Andras Lukacs, David G. Embrick, and J. Talmadge Wright

Emerging new media is a battleground of ideologies and social power. This book was guided by our belief that digital entertainment and video games should not be looked at as a single analytical category. Rather, the many interplaying dialectics among political economy, representations, and audiences must be the foundation of a critical media sociology. While top-down pressures (i.e., economic realities, advertising, marketing, and design conventions) heavily influence new media, emerging new media also responds to challenges from the bottom-up (through countercultures, alternative gaming, consumer demands, to name a few). Our goal with this volume was to show the complexity of this interaction by focusing on three dimensions: social-psychological issues, social inclusions and exclusions, and fandom. We dealt with other important topics such as theories of play, political economy, and methodology in our previously published book, *Utopic Dreams and Apocalyptic Fantasies: Critical Approaches to Researching Video Game Play.*[1]

TECHNOLOGY, POWER, AND LIBERATORY FANTASIES

As we are putting the final touches on this manuscript, the *World of Warcraft* community is fuming over a recent decision by Blizzard Activision to ban (and two days later unban) Swifty, one of its YouTube celebrities, for crashing multiple game servers. As far as Internet sensations go, the "Swifty Riots" might be forgotten by the time this book sees publication. Nonetheless, a closer look at Swifty's turmoil underscores this volume's analysis of the matrix of technology, corporate power, fandom, questions of authority and leadership, and social-psychological implications of modern play.

Swifty became well known in the *World of Warcraft* community through his player-versus-player (pvp) videos on YouTube. Currently his channel has over 189,000 subscribers. The popularity of his videos earned him a sponsorship from Razer USA, a gaming peripherals manufacturer. On July 17, 2011, Swifty was banned for crashing several *World of Warcraft* servers during a live-streamed Razer product giveaway. Almost 5,000 participants were asked to gather in the Alliance city of Stormwind and chant "Swifty Invasion," essentially overloading the chat-server and bringing the realm offline. Blizzard's server technicians responded by immediately banning the leaders of the event.

The next morning, a devastated Swifty posted a short video on YouTube: "I know a lot of you guys probably will say, it's just a game, it's no big deal. But I—I can't say that. It's a game I've been playing for 6 years, I've been playing the same toon for 6 years. It's a part of my life. It's my passion. . . . Sorry guys, trying to figure this out atm. Thx for the support."[2] As expected, outraged Swifty fan invaded the official game forums, and supporters gathered on Swifty's home server chanting "unban Swifty." YouTube videos and blog posts mushroomed debating the merits of Blizzard's action. Swifty's corporate sponsor, Razer USA contacted Bashiok, Community Manager of *World of Warcraft* to help re-activate their client's account. Within 48 hours Swifty was back in the game and on YouTube: "Guys, you have no idea how that feels, this stuff that you guys did for me is unbelievable and I wanna thank you guys. . . . I have never felt such a strong connection with my fans before."[3]

A closer look at the "Swifty Riots" from a critical perspective brings up numerous issues we attempted to address in this book. First, the Internet and social media challenge our traditional understanding of authority, leadership, trust, and solidarity. While J. Talmadge Wright's (chapter 4) student informants occupied virtual realms partially due to the disappearance and commercialization of everyday lived space and relied on existing trust-networks to create a hybrid social world, Swifty's event had a different power-dynamic. The event had a leader and a corporate sponsor, yet it still resembled a horizontal network. Without means of coercion, Swifty was unable to control participants once the spam-snowball started rolling.

Vanessa Long's contribution (chapter 3) helps us to understand the relationship between the player, the screen, the game, and the generation of meaning for game players. The devastation of a banned player or the outrage of fans losing their object of fandom is a stark reminder that games lack fixed meaning and it is up to the players to fill them with value. Game avatars, personal histories, friendship networks, and psychological connections are not disposable or easily replaceable. Bashiok, Community Manager of *World of Warcraft* sarcastically noted that Razer did not send suitcases full of unmarked bills to accelerate or influence the decision to reverse the ban.[4] Blizzard called its action a correction "based on evidence."[5] Elizabeth Erkenbrack's work (chapter 2) prompts us to wonder how various participation frameworks are considered for online policy decisions and what is considered evidence. An event like Swifty's is taking place in a digital gamespace, on a voice-over-IP server, broadcast through a livestream and archived on YouTube. Participants might share the same physical space and use alternative channels to communicate. These laminated discourses compromise the totality of social performance – making clear-cut decisions problematic.

One of the most interesting dimensions of the "Swifty Riots" were the fan reactions and the ensuing protests in-game, on official and fan forums and on YouTube. We are particularly pleased to have a full section of this book dedicated to fan studies. Sean C. Duncan's analysis (chapter 10) between fans and game designers allows us to better conceptualize the discursive strategies players and developers use to frame their activities. Mia Consalvo (chapter 11) broadens the discussion by looking at transnational fandom through Japanese video games. Finally, R. M. Milner (chapter 12) questions the dichotomous approach to producers and fans. As Swifty's fame illustrates, bloggers and modders are in fact productive consumers, sometimes sponsored by gaming companies, and their mostly unpaid work is essential to the operation of contemporary virtual play realms.

While the above-described Swifty episode highlights certain aspect of the matrix of technology, fandom, social psychology, and power relationships, it does not revolve around issues of representation. Nonetheless, representation is still a vital component of a critical media sociology and we devoted a large segment of this volume to these concerns. We invited contributors whose research locates various game narratives and the use of such narratives within larger discourses of power. Joel Ritsema and Bhoomi K. Thakore (chapter 8) showed how popular titles, such as *World of Warcraft* are predicated upon a white, heterosexual masculinity framework and heavily rely on "sincere fictions of whiteness" as an ideological shortcut to establish a meaningful and familiar game-world. David Dietrich (chapter 6) surveyed a broad range of MMPORGs looking at the procedure of avatar creation. He concluded that the lack of racial representations maintains and reinforces normative whiteness in virtual realms. Jessie Daniels and Nick Lalone (chapter 5) noted that systematic racism is not only displayed, but also enacted. Although players often attempt to cloak overt or covert racial remarks as innocent humor or try to minimize their significance to save face, the procedural rhetoric of virtual realms is far from neutral.

Since race, class, and gender inequality constitutes much of the fabric of contemporary society, these representations are not necessarily manifestations of individual or corporate politics, but a mirror image of our transnational capitalist structure itself. Thus, the way developers and artists imagine consumer hardware technology, software applications, art, and game-play elements highlight processes that are beyond the immaterial worlds of virtual realms. For instance, Adrienne L. Massanari (chapter 7) showed the connection between the design of a technological object and ideological motivations: while the Wii platform was successful in recruiting a larger women user bases, its design, marketing, and gameplay is still reliant on taken-for-granted gender assumptions. Of course players can reject and re-interpret the "fixed-meaning" of various representations as Elena Bertozzi (chapter 1) demonstrated with her analysis of traditional masculinity in *Grand Theft Auto*. Yet, Zek Cypress Valkyrie's chapter (chapter 9) is a stark reminder that normative standards of femininity and masculinity (and by extension, race and class ideologies) are still vigorously policed and reinforced within game environments.

SO . . . WHERE DO WE GO FROM HERE?

Like every concluding chapter in a book like ours, we end by addressing the requisite question, "where do we go from here?" Technology and electronic gaming change rapidly, sometimes in the blink of an eye. Yet, our sociological understanding of people and virtual technologies often lag far behind. Part of the reason is because we are just beginning to develop the methodological and theoretical tools needed to engage as researchers in this still new terrain. As scholars, we continue to argue with our Institutional Review Boards over how to properly conduct research on the Internet. And, even as it becomes easier to conduct interviews across the globe, it also becomes increasingly difficult to determine the "real" demographics of respondents as opposed to what may be their alternate selves (i.e., virtual avatars).

Part of the explanation lies in the fact that the social sciences still have not taken seriously existing scholarship in virtual gaming as an object for sociological concern. While extensive research has been undertaken on such topics as social networking (e.g., Facebook), chat rooms, and other internet related research areas, by sociologists in the specialized areas of technology and science, little has been presented or published on digital play and electronic fantasies as a sociological research areas by sociologists, with the exception of work by Sheri Turkle, William Sims Bainbridge, and T.L. Taylor. Most contemporary research on the topic has emerged from departments of Computer Science, Anthropology, Education, Communication, Psychology, English, or Game Studies. Further, research in this area is often marginalized by departments, universities, or even the field itself. One of our objectives in writing this book was to showcase existing, cutting-edge research that has empirical and theoretical relevance to our everyday lives and to the field of sociology in particular. Nonetheless, much more needs to be done. We offer a few questions as to what needs to happen next in order to continue. These questions are not meant to be exhaustive, but merely raise issues which sociology can best inform.

- How has the expansion of the technology of virtual gaming environments altered our concepts of play/fantasy, and imagination? We need to move beyond a critique of the commodity fetish to ask deeper questions about the social role of play, technology, and self-expression.
- What is the impact of digital gaming software on the social processes of solidarity and social exclusion that we find online?
- How should we investigate and understand the relationships between adult players and children? How is authority altered in the virtual environments and what is their potential for empowerment or repression?
- What are the informal social rules established by fans that work to mediate online conflict and facilitate cooperation? How do informal social rules fit with race and gendered forms of social exclusion still practiced in everyday life?
- What might a political economy of digital play look like and what are its relationships with game texts, and audience or fan understandings of play?
- What social science theories can we employ to explore how both cultural representations and player interactions in these digital playgrounds explain transcendence as well as the reproduction of existing social inequalities in the world at large?

In order to better frame these questions, we should think about three areas of consideration: (1) theoretical concerns, (2) methodological concerns, and (3) ethical concerns. Over the past 15 years, much of the theoretical discourse on computer games has been framed by the competing theories of ludology and narratology, that is, play is paramount or stories are paramount in explaining the pleasure of gaming. Both perspectives present a false binary, as if play and storytelling were separate affairs. Aside from this theoretical reduction, the empirical reduction present in extensive textual analysis and survey work often ignores larger issues of social power, whether they are online or offline. In addition, the domination of psychological models reduces the significance of digital play to one of individual psychology and satisfaction, rarely asking broader and more critical questions of the origins of desire itself and how that desire fits into the authority of any given society. While it is tempting to bring in symbolic interaction as the primary model of social theory to explain in-game behavior, we should not limit ourselves to this and instead look at a more critical approach without ignoring the pleasure involved in game inter-actions.

Methodologically, studying the "play" of computer games, while initially stuck in elaborate textual analysis of game content, has broadened to include extensive online survey work, participant observation studies, and inter-views, both online and offline. We encourage the expansion of longitudinal studies as in the work of Dmitri Williams, as well as the participant observa-tions of T.L. Taylor, Bonnie Nardi, Mia Consalvo, Celia Pearce, Tom Boell-stroff, and others. The debate over whether we should consider the virtual world as "real" in its own right or simply a representation of the real is a debate which must occur in sociology, as it has in anthropology.

Finally, ethical considerations are of key concern to us. We continue to examine the question of informed consent in a virtual world and ask, what are ways we can do this research without causing harm to others? How do license requirements and confidentiality influence the results we obtain in our research? When and how does one reveal their identity to others in a virtual realm where you are represented by a self-created avatar? And, final-ly, what are the real possibilities for harm, and what are the fears which need to be assuaged for such research to gain legitimacy?

EMBRACING THE FUTURE OF DIGITAL GAMING

While we attempted to include analyses about various games, virtual realms, and gaming platforms in this volume, our goal was not to provide a compre-hensive list of contemporary games. Individual chapters contain vivid de-scriptions and snapshots of certain titles, and particular ethnographic details add considerable depth and value to this book. Nonetheless, we approached various titles and franchises as metaphors, windows that allow us to observe and understand more about the complexity of today's interactive media land-scape that spans beyond individual titles, technological platforms, and media applications.

Pierre Bourdieu, when talking about the symbolic power of proper lan-guage, noted that "the game is over when people start wondering if the cake is worth the candle."[6] We often feel that the dichotomizing power of contem-porary debates about the role of media forces scholars to either abandon their critical approach to become defenders of digital entertainment, or fail to appreciate the benefits and blessings of new media amidst their zealous at-tacks on video games. We hope that the chapters of this volume demonstrat-ed that media scholarship can embrace the dialectic of social gaming. This dialectic revolves around for-profit enterprises developing titles by borrowing, transplanting and transforming dominant ideologies, representations, and stereotypes. Packaged products are marketed under the rules of the transnational capitalist system. Yet, the end-user is never a passive consumer of media products and the ideologies contained within. Agentic users, or productive consumers to use Milner's term, form communities of meaning, mesh and modify content, create a lively fan culture, and take ownership of virtual realms by challenging operators and publishers.

Digital gaming, communication technology, and virtual spaces saturate the everyday world at an astonishing speed, often violating conventional rules and traditions, sometimes even turning the rules of established dissent upside down. We hope that the contribution of critical media scholarship, and researchers who are willing to embrace the messy and contradictory nature of new technology will provide a good starting point for policy debates, future research, and even inspire game designers to keep social justice and liberation in focus while developing new titles.

NOTES

1. J. Talmadge Wright, David G. Embrick, and Andras Lukacs, eds., *Utopic Dreams and Apocalyptic Fantasies: Critical Approaches to Researching Video Game Play* (Lanham, MD: Lexington Books, 2010).
2. johnsju, "Swifty Banned" (YouTube.com, June 18, 2011), accessed at http://www.youtube.com/watch?v=J-DFfBUA6vM.
3. johnsju, "Swifty Banned Update" (YouTube.com, July 20, 2011), accessed at http://www.youtube.com/watch?v=N3-jS_nYY4s&feature=relmfu.
4. Nonetheless, the connection between Activision Blizzard and Razer USA prompts questions about corporate interest and player equality under ELUAs. Are gaming celebrities afforded special privileges because of their marketing-work and ability to mobilize fans?
5. Bashiok, "(Locked) Actions for Recent Realm Disruptions," (Blizzard Entertainment, July 19, 2011), accessed at http://us.battle.net/wow/en/forum/topic/2842767666.
6. Pierre Bourdieu, *Language & Symbolic Power* (Cambridge, MA: Harvard University Press, 2003), 58.

Index

hotfix, 184
House of the Dead 2, 56, 61n31
Howard, Todd, 229
Huizinga, Johann, 18, 177
humor: Internet and, 95; and racial issues, 95–96
hunting, and masculinity, 9–11
hyper-resonance, 160–161; as preferred, 160, 161; term, 160, 173n27

id, 42
identification: with avatars, 43, 46, 48; Lacan on, 42; video game play and, 43
identity tourism, 85, 93; versus hyper-resonance, 160; term, 16
Imaginary, 49
immigrants, 92
imported games, 211–212
Indiana Jones films, 17
in-game activity, 23; term, 25
in-room speech, 29, 33–37
instant gratification, 45
insults, trust and, 74
intelligence: *Grand Theft Auto* and, 7; masculine films and, 6; and masculinity, 9–11
intermediate action: in game types, 55–57; types of, 51–57
intermediate ego, 41–62n33; nature of, 44–47; space of, 48–49
Internet: and humor, 95; and race, 87
Internet cafes, 69, 70, 70–71, 80
interpersonal alignment, in *World of Warcraft*, 28–30, 35–37
invisibility, 87, 149
Iron Man, 6, 17
irony, *Grand Theft Auto* and, 19

Japan, travel to, 214–215
Japanese video games, 199–219n29; attractions of, 206–208; introductions to, 204–208
Jenkins, Henry, 64, 120, 201, 223, 224
Jillian Michaels' Fitness Ultimatum, 127
Joystiq, 124
juiciness: in casual games, 128; in *Wii Fit*, 129
Juul, Jesper, 64, 117, 122, 128

Katamary Damacy, 207
Kendall, Lori, 87
Kingdom Hearts, 46, 61n15, 205
King of Fighters 2002 Unlimited Match, 212
Klaus, Elisabeth, 126
Kline, Stephen, 71
Korean players, 102
Ku Klux Klan, 90

labor: fans and, 221–222, 225–226; New Organization and, 227; symbolic, 149
Lacan, Jacques, 42, 44, 49
lack, psychoanalytic, 44–45; confrontation with, 46–47
lag, 102; term, 115n27
Lalone, Nick, 85–99n51
language: on culture, difficulty with, 208–211; Japanese, 213–214. *See also* discourse analysis
Laurel, Brenda, 120
Law, Jude, 118
leisure, fans and, 221–222
Lewis, C. S., 144
literacy, 79
literature, fantasy, racial elements in, 144–145
location. *See* place
LOLcats, 95
Long, Vanessa, 41–62n33; online game, 108, 109, 146; racial issues in, 144–145
ludic play, 180; in online affinity space, 177–198n16
Lukacs, Andras, xi–xixn5, 247–253n6
Luna Online, 109
Lunar Silver Star Story, 155, 171n2

machinima, 226
Madden, 171n3, 221, 242n1
mainstream games, racism in, 91–93
manga, 203, 206
Mario and Sonic at the Olympic Winter Games, 127
masculinity: changing images of, 4, 5; *Counter-Strike* and, 74–75; films and, 4, 5–8, 118; *Grand Theft Auto* and, 3–22n57; models of, 4; performance of, 15; technology and, 5
mashups, 223, 243n17